CURRENT AT THE
NANOSCALE
An Introduction to Nanoelectronics

T0344258

CURRENT AT THE
NANOSCALE

An Introduction to Nanoelectronics

Colm Durkan

University of Cambridge, UK

Imperial College Press

Published by

Imperial College Press
57 Shelton Street
Covent Garden
London WC2H 9HE

Distributed by

World Scientific Publishing Co. Pte. Ltd.
5 Toh Tuck Link, Singapore 596224
USA office: 27 Warren Street, Suite 401-402, Hackensack, NJ 07601
UK office: 57 Shelton Street, Covent Garden, London WC2H 9HE

British Library Cataloguing-in-Publication Data
A catalogue record for this book is available from the British Library.

For photocopying of material in this volume, please pay a copying fee through the Copyright Clearance Center, Inc., 222 Rosewood Drive, Danvers, MA 01923, USA. In this case permission to photocopy is not required from the publisher.

ISBN-13 978-1-86094-823-7
ISBN-10 1-86094-823-5

Printed in Singapore.

For Biddy, Ben and Rosie

Preface

We are in an age where, in the Western world at least, we all take high technology pretty much for granted. Many of us use and even depend on mobile phones, laptops, PDAs and other miniature electronic devices during our daily lives. What has made this all possible are advances in semiconductor technology, at the centre of which is the humble transistor. Next to the discovery of penicillin, the invention of the solid-state transistor was arguably one of the most important developments for mankind over the past century, as it has had far-reaching consequences across all aspects of life.

As the scale of electronic devices continues to decrease, and approaches the nm level, it is becoming more and more important to understand the details of current flow in reduced dimensions. This book is intended as an introductory overview of transport phenomena from the macroscale right down to the atomic level. There are two means by which nm-scale devices can be fabricated — by top-down or bottom-up technologies. The top-down approach has been used with obvious success by the semiconductor industry for several decades now and the bottom-up approach is being pioneered by researchers in the field of nanotechnology. Whilst current indications are that bottom-up nanotechnologies will not completely replace top-down technologies, the two will almost certainly become complementary. In any event, it will still be essential to have a sound understanding of transport at the nanoscale. We begin in Chapter 1 by looking at transport in familiar devices: resistors and transistors, and how, despite any ideas we may have to the contrary, their electrical characteristics are determined by events at the quantum level. We then dedicate Chapter 2 to gaining a practical understanding of the quantum nature of current flow, i.e. the relationship between current and voltage and the origins of electrical resistance. In Chapter 3, we look at the boundary between the quantum and macroscopic regimes — known as the mesoscopic regime, and look at how geometry, size and microstructure start to play an important role in determining resistance at the nanoscale. In Chapter 4, we look at the techniques used to probe the electrical properties of structures and devices at the nanoscale — scanning probe microscopy. In fact, it was the development of scanning probe microscopes that jump-started the field of nanotechnology. In Chapter 5, we look at some of the detrimental effects of current flow: heating and electromigration in nanowires. This is particularly important due to the fact that while transistors in microprocessors continue to shrink, so too must the interconnects which join them together. The resilience of small wires to the flow of current for prolonged periods is not the same as for large (i.e. micron-scale) wires. In the final chapter, we take a look at the field of molecular electronics, which has gained a lot of interest in recent years due to the promise it holds for novel circuit functions. This text is intended to serve as a useful introduction to quantum mechanics, scanning-probe microscopy and electronic transport.

C. Durkan
Cambridge, January 2007

Contents

Chapter 1

Macroscopic Current Flow

In order to be able to gain any insight into current flow at the nanoscale, we must first consider what happens at macroscopic scales and then see the effect of reducing the dimensions to the nanoscale. The behaviour of conventional electrical circuits can be understood in quite simple terms, whereas at the nanoscale, there are a number of subtle effects which can only be understood within the framework of quantum mechanics, as we shall see in this chapter and in Chapter 2. In between these disparate regimes, we have *mesoscopic* transport, which we will briefly consider in Chapter 3.

On the basis of his detailed experimental observations in the 1820s, Georg Ohm formulated his famous Law [1] which states the following:

For a constant temperature, the current flowing through a conductor is directly proportional to the potential difference between its ends. Or, V = IR.

This is illustrated in Fig. 1.1. The constant of proportionality between voltage and current is known as resistance, R. The resistance in turn depends on the geometry of the conductor and a material constant, resistivity (ρ) as $R = \rho l/A$, where l and A are the length and cross-sectional area of the conductor, respectively. As we will see later, when any of the dimensions of a conductor are at the nanoscale, resistivity itself becomes dependant on geometry. For now however, let us consider Ohm's law in more detail, from a purely classical (i.e. non-quantum) standpoint — Drude's model of electronic conduction.

We will then introduce the necessary corrections which must be made to be consistent with quantum mechanics. We would like to explain the following points:

1. What is electric current?
2. Why and how does current depend on voltage?
3. What are typical values of resistance/resistivity of conductors?
4. What is the effect of changing temperature?

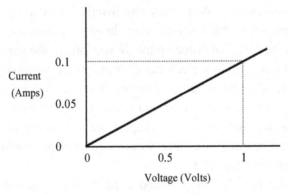

Fig. 1.1. Relationship between current and voltage for a conductor. In this case the resistance is 10 Ω.

1.1. The Classical (Drude) Model of Electronic Conduction and Ohm's Law

To answer the above questions, we need to consider what happens when we apply a voltage across a conductor. The voltage causes there to be a local electric field E, within the conductor. This electric field induces a force on the electrons (of charge, $e = -1.6 \times 10^{-19}$ C) of strength eE, in a direction opposite to that of the applied field. On this basis, we would therefore expect that electrons will continue to accelerate as they traverse the conductor. The flow of charge carriers within a conductor is known as a current, and the magnitude of the current is the amount of charge (in Coulombs) passing a point in the conductor per second. A current of 1 A corresponds to 3.25×10^{18} electrons passing a point per second.

In reality, we know that electrons do not accelerate indefinitely as they flow through a conductor, but they drift along at a finite speed (called the drift velocity, typically 10^6 m.s^{-1}). This drift velocity is somewhat analogous to the terminal velocity experienced by falling objects (which continually loose momentum to air molecules), and is due to the electrons colliding with impurities, lattice imperfections and lattice vibrations (phonons) within the conductor. The average distance and time between collisions are the *mean free path, ℓ and mean free time, τ* respectively and are of the order 10–50 nm and 10^{-14} s for a metal at room temperature [2]. Given that the force on an electron is eE, and force is mass times acceleration, we find that the average velocity (*acceleration × time*) of the electrons immediately before a collision is $v = eE\tau/m$. If we have n electrons per unit volume, then it follows that the current density, $J = nev = ne^2 \tau E/m$. This is essentially a statement of Ohm's law, as it predicts that the current density is proportional to the electric field, which will be proportional to the applied voltage. The constant of proportionality between current density and electric field is the conductivity, σ; i.e. $J = \sigma E$, where $\sigma = ne^2\tau/m$. The electrical resistivity, ρ is $1/\sigma$, so $\rho = m/ne^2\tau$.

To investigate the effect on the current of changing temperature, we need only look at the above formula. As we increase temperature, we cause the atoms in the conductor to move more vigorously. This has the effect of reducing τ and ℓ, and consequently increases the resistivity. In practice, as we increase the current flowing through a conductor, we also increase the number of collisions between the electrons and entities within the conductor, which has the effect of heating the conductor. This is the principle behind the operation of electric filament heaters and light bulbs. There comes a point when this heating causes the resistance to increase, and the current–voltage characteristics become non-linear, instead following the curve shown in Fig. 1.2 (also see Chapter 5).

Thus far, whilst we have deduced Ohm's law, which is an experimental fact, we have not considered the quantum nature of materials. In order to gain a deeper insight into conduction, we must now consider the free-electron model of electronic conduction.

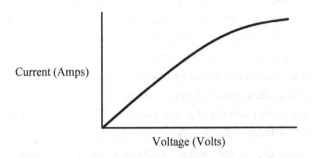

Fig. 1.2. Relationship between current and voltage for a conductor, where the current induces heating.

1.2. The Quantum (Free-Electron) Model of Electronic Conduction

In the Drude model which we have just discussed, we have implicitly assumed the following:

- The electrons do not interact with each other or the crystal lattice;
- The electrons can have *any* velocity and hence, energy;
- All electrons within a conductor contribute to conduction;
- The electron gas follows Maxwell–Boltzmann statistics.

The first assumption can be justified for most metals, where the atomic potential and electronic interactions are effectively screened out due to the high electron density. For semiconductors however, the situation is altogether different, leading to the formation of a band-gap.

The concept of the electrons inside a conductor not interacting with anything is analogous to a gas, hence the name *free-electron gas*. The second assumption is not viable, as quantum mechanics tells us that the energy of an electron in a conductor will have certain, discrete values. Finally, electrons are Fermions, and are known not to follow

Maxwell–Boltzmann statistics, but rather Fermi–Dirac statistics. These facts ultimately lead to the failure of the Drude model. The Drude model simply cannot explain a large number of phenomena, e.g. the specific heat of materials, the resistivity of materials, and many other points which we will visit later. To see where this discreteness comes from, we need to turn to a quantum description of the electrons in a conductor. Consider a cubic-shaped conductor, of side, L, as shown in Fig. 1.3. The time-independent free-particle Schrödinger equation is

$$-\frac{\hbar^2}{2m}\left(\frac{\partial^2}{\partial x^2} + \frac{\partial^2}{\partial y^2} + \frac{\partial^2}{\partial z^2}\right)\psi_k(x,y,z) = E_k\psi_k(x,y,z). \qquad (1.1)$$

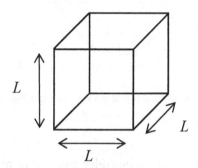

Fig. 1.3. Cube of conductor, side L, used for calculating the number of electron states.

The solution to this is

$$\psi_k(x,y,z) = A\sin(l\pi x/L)\sin(m\pi y/L)\sin(n\pi z/L), \qquad (1.2)$$

where l, m, and n are integers. For generality, we must satisfy the periodic boundary condition (known as the *Born von Karman* boundary condition) that $\psi_k(x+L,y,z) = \psi_k(x,y,z)$, and similarly for y and z. This condition can only be met when l, m and n are all *even* numbers, so instead of having $k_x = l\pi/L$, we actually have $k_x = 2l\pi/L$, and similarly for k_y and k_z.

The energy of a given wavefunction (state) is therefore $E_k = \hbar^2 k^2/2m$ (i.e. Energy $= (1/2)mv^2$), giving a parabolic E_k–k (dispersion) relationship, as illustrated in Fig. 1.4.

Substituting for the values of k from above, we get

$$E_k = 4\pi^2 \hbar^2 (l^2 + m^2 + n^2)/2mL^2, \qquad (1.3)$$

which clearly has discrete values — a defining characteristic of quantum systems.

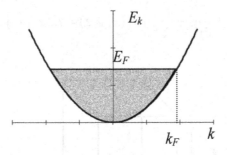

Fig. 1.4. Energy (E_k)–momentum (k) dispersion relationship according to the free-electron model.

As an example, let us consider two cubic pieces of conductor with different dimensions; one has a side of 5 nm, the other is 1 mm. The spacing between the first and second levels is given by

$$\Delta E_k = 3h^2/2mL^2. \qquad (1.4)$$

For the 5 nm conductor, this corresponds to 0.179 eV, whereas for the 1 mm conductor it is 4.5×10^{-12} eV. Given that electrons have a thermal energy of $k_B T$, which is around 25 meV at room temperature, we will not notice energy quantization in macroscopic conductors, but once any of the dimensions of the conductor are of the order 10 nm, it will have a marked effect. Figure 1.5 is a plot of the $\Delta E_k/k_B T$ as a function of size, where we have highlighted the transition between quantum, mesoscopic and macroscopic regimes. The boundaries between these regions will shift to larger feature sizes as the temperature is reduced, and if we use semiconductors, as the electrons then have a lower effective mass.

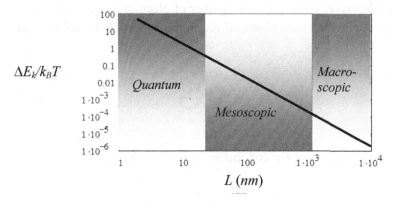

Fig. 1.5. Plot of $\Delta E_k / k_B T$ versus conductor size, for conductors ranging between 1 nm and 10 μm across.

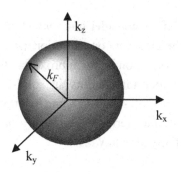

Fig. 1.6. Fermi sphere. All the k states up to k_F are filled.

As there are a finite number of electrons within a conductor, all the energy states up to a particular value (the *Fermi energy, E_F*) will be filled, as shown in Fig. 1.4. If we consider a single electron travelling through the conductor, it will have a wave-function of the form $\psi_k(r) = e^{ik \cdot r}$. It is instructive to think of the occupied states as being represented by points within a sphere in k-space. The surface of the sphere, where $k = k_F$ is known as the *Fermi surface*, and represents states at the Fermi energy, as illustrated in Fig. 1.6.

The energy of electrons at the Fermi surface is just $E_F = \hbar^2 k_F^2/2m$. In order to calculate the value of k_F, we note that there are only two points for each volume element $(2\pi/L)^3$ (the factor of 2 comes from two possible values of spin). The number of electrons in the conductor, N, is the ratio of the total volume of the Fermi sphere to the volume per state, i.e.

$$N = 2 \frac{4\pi k_F^3/3}{\left(2\pi/L\right)^3} = \frac{V k_F^3}{3\pi^2}, \tag{1.5}$$

where V is the volume of the conductor. It follows that

$$k_F = \left(\frac{3\pi^2 N}{V}\right)^{1/3} \Rightarrow E_F = \frac{\hbar^2}{2m}\left(\frac{3\pi^2 N}{V}\right)^{2/3}. \tag{1.6}$$

As an example, if we consider gold, which has a Fermi energy of 5.51 eV, we find that the electron concentration $(n = N/V)$ is 5.93×10^{28} electrons.m^{-3}, which corresponds to approximately one electron per atom. An extremely important quantity in solid-state physics, and in nanoscience is the number of states per unit energy within the energy interval dE, per unit volume, termed the *density of states*, $\mathcal{D}(E)$ of a material. This is essentially $(1/V)dN/dE$.

In 3D, the volume of a shell, radius k, thickness $dk = 4\pi k^2 dk$

$$\text{Volume per } k\text{-state} = \left(\frac{2\pi}{L}\right)^3$$

$$\Rightarrow \text{Density of states} = 2\left(\frac{4\pi k^2 dk}{\left(\frac{2\pi}{L}\right)^3}\right) = \frac{L^3 k^2 dk}{\pi^2}$$

$$E = \frac{\hbar^2 k^2}{2m} \Rightarrow k = \sqrt{\frac{2mE}{\hbar^2}} \Rightarrow dk = \left(\frac{2mE}{\hbar^2}\right)^{-\frac{1}{2}} \frac{m}{\hbar^2} dE$$

$$\Rightarrow \mathcal{D}(E)_{3D} = \frac{E^{1/2}}{2\pi^2} \left(\frac{2m}{\hbar^2}\right)^{3/2}. \tag{1.7}$$

In *two* dimensions (where there is confinement in one direction — a *quantum well*), the Fermi sphere is replaced by a Fermi circle, and instead of calculating the number of *k*-states in a shell of radius *k*, thickness *dk*, we need to calculate the number in an annular ring of radius *k*, thickness *dk*, which is just $2\pi k dk$. The area per state in 2D is $(2\pi/L)^2$. Thus, the 2D density of states, taking into account spin degeneracy is $kdkL^2/\pi$, or in terms of number of states per unit area, is kdk/π. In terms of energy, this is

$$\mathcal{D}(E)_{2D} = k\frac{dk}{\pi} = \sqrt{\frac{2mE}{\hbar}} \left(\frac{2mE}{\hbar^2}\right)^{-\frac{1}{2}} \frac{m}{\hbar^2} dE$$

$$= \frac{m}{\pi\hbar^2} dE. \tag{1.8}$$

Now, this is the density of states for each value of bound state energy within the quantum well. The total density of states, taking into account all of the bound states is:

$$\mathcal{D}(E)_{2D} = \frac{m}{\pi\hbar^2} \sum_i H(E - E_i) dE, \tag{1.9}$$

where $H(E - E_i)$ is the Heaviside step function.

In *one* dimension, i.e. a *quantum wire*, the density of states is $2Ldk/\pi$, or per unit of space is $2dk/\pi$. This gives:

$$\mathcal{D}(E)_{1D} = 2\frac{dk}{\pi} = \frac{2}{\pi}\left(\frac{2mE}{\hbar^2}\right)^{-\frac{1}{2}}\frac{m}{\hbar^2}dE$$

$$= \frac{1}{\pi}\left(\frac{m}{\hbar^2}\right)^{\frac{1}{2}}\frac{1}{E^{\frac{1}{2}}}dE. \qquad (1.10)$$

These have the form of peaks at each of the quantized energy levels, and are known as *van hove singularities*.

A consequence of this is that there is a *very* strong dimensional dependence on the electrical properties of conductors, as we shall see later.

Thus far, we have neglected the effect of temperature, and our analysis is only valid for $T = 0$ K. As we increase temperature, the kinetic energy of the electrons within the conductor will increase, with the effect of smearing out the Fermi surface. This happens as some of the electrons at and just below the Fermi energy will go into slightly higher energy states just above the Fermi energy, depopulating their original states. The net result of this is that, around the Fermi energy, the probability of a state being occupied is no longer a simple step function (1 below E_F, 0 above E_F), but is determined by the *Fermi–Dirac distribution*, $f(E)$. The details of the derivation of this function have been reported in numerous solid-state and statistical mechanics textbooks, and we are only concerned with the result, which is

$$f(E) = \frac{1}{e^{\left(\frac{E-\mu}{k_BT}\right)}+1}, \qquad (1.11)$$

where μ is the chemical potential of the conductor: the energy of the highest occupied state, which is E_F at absolute zero. The product of $f(E)\mathcal{D}(E)$ is the density of *filled* states, and is shown in Fig. 1.7.

From a quantum mechanical standpoint, how does conduction occur? We have already ascertained that the electrons within the conductor experience a force eE. Force is rate of change of momentum, which in quantum mechanical terms is the rate of change of $\hbar k$, i.e. $eE = d(\hbar k)/dt$. Thus, the change in the k-vector in a time, t is $eEt/\hbar$. If we assume that after each collision between the electron and the conductor, the velocity goes to zero, then the velocity gained in one mean free path is $\hbar\Delta k/m = eE\tau/m$, the same result we obtained earlier. If we apply an electric field in the $-x$ direction, the Fermi sphere moves in the $+x$ direction, eventually maintaining a steady position due to scattering events. This is illustrated in Fig. 1.8.

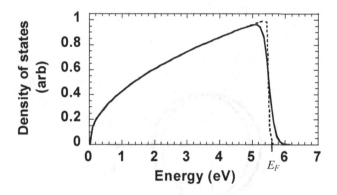

Fig. 1.7. Density of occupied states for gold ($E_F = 5.51$ eV). The solid and dotted curves are for $T = 1000$ K and 20 K, respectively.

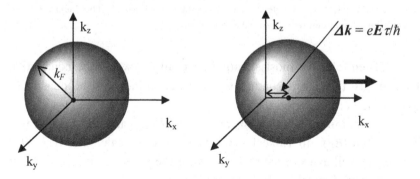

Fig. 1.8. Fermi sphere in the presence of an electric field.

We can better appreciate the subtleties of electronic transport if we look at a cross-section through the Fermi sphere, as shown in Fig. 1.9.

In summary, when we apply a voltage across a conductor, the electrons acquire some momentum opposite to the direction of the electric field. For many of the electrons within the conductor, this additional momentum is less than the random momentum they already have ($\hbar k$), and the only electrons which contribute to conduction are those which are near the leading edge of the Fermi surface, which is the shaded region shown in Fig. 1.9.

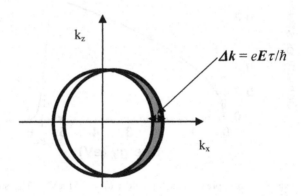

Fig. 1.9. Slice through the Fermi sphere in the presence of an electric field. The dark shaded region shows the new states around the Fermi energy which have been populated due to the field. These are the electrons contributing to conduction. The thickness of the line describing the circumference of the Fermi sphere is $k_B T$.

Therefore, for most situations, we might expect the conduction of electricity through a conductor to be determined by the density of states at the Fermi energy, $\mathcal{D}(E_F)$.

Combining the equations above for $\mathcal{D}(E)$ and E_F, we find that $\mathcal{D}(E_F) = 3n/(2E_F)$. In Table 1.1, we have listed the calculated n, E_F and $\mathcal{D}(E_F)$, as well as experimental values for the conductivity for some of the most technologically relevant metals.

The disparity between the calculated and observed trends in $\mathcal{D}(E_F)$ versus σ illustrates that conduction is not just determined by the number of electrons within the conductor, but there are other factors at play.

To gain an understanding of this, we need to consider a more detailed version of conduction through materials — the *nearly-free-electron model*. Only then will we be able to appreciate the dominant factor in determining the electrical resistance of a conductor — scattering. This will then lead us onto phonons and the effect of temperature on electrical resistance.

Table 1.1. Calculated conduction parameters of selected metals.

Metal	Number of electrons/m^3	Fermi energy, eV	$\mathcal{D}(E_F)$, arb. units	σ (measured) $10^6\,\Omega^{-1}.\text{cm}^{-1}$
Au	5.90×10^{28}	5.51	1.07	0.45
Ag	5.85×10^{28}	5.48	1.07	0.63
Cu	8.45×10^{28}	7	1.21	0.59
Al	18.06×10^{28}	11.63	1.55	0.38

1.3. The Nearly-Free-Electron Model of Electronic Conduction and Band Structure

In crystalline materials, the electrons are in a periodic potential. The form of the electrical potential in a crystal is such that there are potential wells centred on the atomic cores. Each of these wells will have discrete allowed energy levels of a form similar to those of Hydrogen. Due to the proximity of the atoms to each other, the tails of the potentials overlap and modify the overall potential. This coupling causes the energy levels to shift and split, and for a number N of atoms, we will have N energy states, corresponding to $2N$ possible electron states (two comes from the two spin states of an electron).

Our aim is to find the form of the E–k relationship for electrons in such a potential, and see how it compares to the free electron case. Then we will be in a better position to understand the origin of the differences between metals, semiconductors and insulators.

There are a number of methods which may be used to investigate the nearly-free-electron model, including the Kronig-Penney model [3], which is an approximate method. We are going to use a more general method, based on solving Schrödinger's equation for an arbitrary periodic potential. Our starting point is to look at the basic form of the potential, as shown in Fig. 1.10.

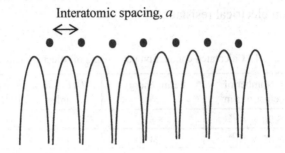

Interatomic spacing, *a*

Fig. 1.10. Periodic potential as seen by an electron in a crystal. The black dots represent atoms.

The steps we will follow are:

1. Define potential as seen by an electron;
2. Expand as a Fourier series;
3. Solution to Schrödinger's equation contains information about periodic potential super-imposed on free-electron wave-functions (which are simple plane waves);
4. This periodic part to the solution can also be expressed as a Fourier series;
5. Insert all of the above into Schrödinger's equation and solve.

The reciprocal lattice vectors are given by *G*, and the lattice spacing is *a*. As we are dealing with a periodic system, it is useful to use the Fourier expansion of the crystal potential, i.e.

$$V(x) = \sum_p V_p e^{iG_p x}, \qquad (1.12)$$

where the Fourier coefficients are given by

$$V_p = \frac{1}{a} \int_0^a V(x) e^{-iG_p x} dx \qquad (1.13)$$

and $p = 0, \pm1, \pm2,\ldots$ and $G_p = 2\pi p / a$.

The general solution to the Schrödinger equation with a periodic potential is $\psi(x) = e^{ikx} u(x)$. This is a plane wave modulated by the function $u(x)$, where $u(x)$ is a periodic function with the periodicity of the lattice, i.e. $u(x)$ represents the influence of the crystal potential. This is known as **Bloch's theorem** [4], and $u(x)$ as a **Bloch function**. In Fig. 1.11, we show the typical form of the wave-functions for the free- and nearly-free-electron models, and we include the approximate lattice potential for reference.

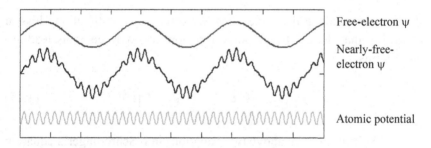

Free-electron ψ

Nearly-free-
electron ψ

Atomic potential

Fig. 1.11. Approximate form of single-particle wave-functions for the free- and nearly-free-electron models, and the atomic potential we are using in the nearly-free-electron model.

In the same way as we expanded the potential as a Fourier series, we can now do the same for $u(x)$, to obtain:

$$u(x) = \sum_n C_n e^{iG_n x}, \qquad (1.14)$$

where $n = 0, \pm1, \pm2, \ldots$ and $G_n = 2\pi n / a$.

That gives for the total expansion of the wave-function:

$$\psi(x) = \sum_n C_n e^{\left(i(k+G_n)x\right)}. \tag{1.15}$$

If we now insert the Fourier expansions of both $\psi(x)$ and $V(x)$ into Schrödinger's equation, we end up with a set of simultaneous equations in the unknown C_n. Note that the coefficients V_p are known, as the form of the crystal potential is assumed initially. There are an infinite number of terms, so to make the problem manageable, we will artificially truncate the series and consider only the leading-order terms given by $n = 0, \pm1$. This is justified for weak potentials such as those found in metals, i.e. we write $V(x)$ as:

$$V(x) = V_0 + V_1 e^{iG_1 x} + V_{-1} e^{iG_{-1} x} = V_0 + 2V_1 \cos(G_1 x). \tag{1.16}$$

If we continue along the same lines, we can assume without any loss of generality that the electronic wave-function also only contains leading-order terms, i.e.

$$\psi(x) = \left[C_0 + C_1 e^{iG_1 x} + C_{-1} e^{iG_{-1} x} \right] e^{ikx}. \tag{1.17}$$

Substituting the above two equations into Schrödinger's equation we find

$$\left(-\frac{\hbar^2}{2m} \frac{d^2}{dx^2} + V_0 + V_1 e^{iG_1 x} + V_{-1} e^{iG_{-1} x} \right) \left[C_0 + C_1 e^{iG_1 x} + C_{-1} e^{iG_{-1} x} \right] e^{ikx}$$

$$= E\left(C_0 + C_1 e^{iG_1 x} + C_{-1} e^{iG_{-1} x} \right) e^{ikx}. \tag{1.18}$$

If we just consider a region where C_0 and C_{-1} dominate (i.e. we are only considering electrons travelling in the positive x-direction, with positive k), we are left with the relationships (noting that $G_{-1} = -G_1$ etc.):

$$\left(\begin{array}{l} -\dfrac{\hbar^2 k^2}{2m} C_0 + V_0 C_0 + C_0 V_1 e^{iG_1 x} + C_0 V_{-1} e^{iG_{-1} x} \\[2mm] -\dfrac{\hbar^2 C_{-1} e^{iG_{-1} x} (k + G_{-1})^2}{2m} + V_0 C_{-1} e^{iG_{-1} x} + V_1 C_{-1} + V_{-1} C_{-1} e^{2iG_{-1} x} \end{array}\right)$$

$$= EC_0 + EC_{-1} e^{iG_{-1} x}. \tag{1.19}$$

Collecting terms in $e^{iG_{-1} x}$, we find that:

$$C_0 V_{-1} = C_{-1}\left(\frac{\hbar^2}{2m}(k + G_{-1})^2 + E - V_0\right). \tag{1.20}$$

Terms without any exponent give:

$$C_{-1} V_1 = C_0\left(\frac{\hbar^2 k^2}{2m} + E - V_0\right). \tag{1.21}$$

For a non-trivial solution, both ratios for C_{-1}/C_0 should be equal, i.e.

$$\left(\frac{\hbar^2 k^2}{2m} + E - V_0\right)\left(\frac{\hbar^2 (k + G_{-1})^2}{2m} + E - V_0\right) = V_1 V_{-1} = |V_1|^2. \tag{1.22}$$

This gives us a solution for positive values of k, and we can follow similar steps as above to include C_0 and C_1 to have solutions with negative values of k. A plot of E versus k is in Fig. 1.12. This is known as a *dispersion relation*, or a *band diagram*.

Points of interest to note relative to free-electron case:

- The energy is shifted up by the amount V_0, where V_0 is the spatial average of $V(x)$.
- At certain values of k, i.e. at $k = -G_{-1}/2 = \pi/a$, gaps appear in the dispersion relationship, which define the *Brillouin Zone* boundary, and this corresponds to Bragg reflection. At this value of k, there are two possible values of energy:

$$E = V_0 + \hbar^2 (\pi/a)^2 \pm |V_1|. \tag{1.23}$$

We can interpret this as the lower and higher energies being the valence and conduction-band edges, respectively. The separation in energy, i.e. the *band-gap* is $2|V_1|$, which is just twice the first term in the Fourier series expansion of the crystal potential, and the Schrödinger equation has no wave-like solutions in this gap.

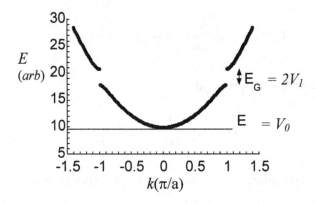

Fig. 1.12. E_k–k (dispersion) relationship according to the nearly-free-electron model.

The wave-functions at the valence- and conduction-band edges are:

$$\psi_{c,v}(x) = C_0 \left(1 \pm e^{-\frac{2i\pi x}{a}} \right) e^{\frac{i\pi x}{a}}, \tag{1.24}$$

which can be written as:

$$\psi_v(x) = 2C_0 \cos\left(\frac{\pi x}{a}\right) \quad \text{and} \quad \psi_c(x) = 2iC_0 \sin\left(\frac{\pi x}{a}\right). \tag{1.25}$$

These are standing waves which means electrons at these energies do not travel through the crystal, and hence do not contribute to conduction.

For each term we add to the Fourier expansion of the crystal potential, there will be a band-gap, at successive multiples of $k = \pi/a$. Real materials tend to have many bands, as the atomic potential is not a simple sine function. Band structure is also a function of the direction we are looking in, as the interatomic spacing will appear different in different directions. A proper calculation would involve all three dimensions as well as the correct form of the atomic potential. The band-gap will depend very strongly on the interatomic distance — the closer together the atoms are, the larger will be the band-gap, as the overall interaction will be stronger. This is evident by looking at the band-gaps of diamond and silicon, which are both formed from group IV elements, C and Si, the only difference being the atomic size. In diamond, the band-gap is over 5 eV, whereas in Si, it is around 1.14 eV. This is directly related to the interatomic spacing which is 235 pm in Si and 142 pm in diamond.

The representation of E versus k in Fig. 1.12 is known as the *extended-zone scheme*. A more-commonly used representation is the *reduced-zone scheme*, as illustrated in Fig. 1.13. Here, the bands are all shifted laterally to enable them to fit within the first Brillouin zone.

Looking at Fig. 1.13, the bands are symmetric about $k = 0$; i.e. the conduction band minimum occurs at the same k-value (in this case, 0) as the valence band maximum. This is known as a *direct-gap* material. There are also *indirect-gap* materials, of which silicon is the most common example. Once the band-gap is too large for electrons to be thermally promoted from the valence-band to the conduction-band, the material is said to be insulating. This occurs once the band-gap is above around 3 eV (this also explains why many insulators are either white or transparent — they do not absorb much light).

So far, we have seen that classical mechanics gets it partly right, that is, it predicts the same value as quantum mechanics does for the conductivity of materials. By employing the nearly-free-electron model, we have demonstrated that we create a band-gap simply by placing electrons in a crystalline material, i.e. one with a regular (periodic) atomic spacing. The band-gap depends on the interatomic spacing and on the atomic potential.

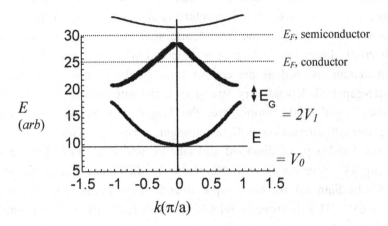

Fig. 1.13. E_k–k (dispersion) relationship according to the nearly-free-electron model, drawn in the reduced-zone scheme, in this case with three bands. The horizontal lines indicate the position of the Fermi energy in a typical conductor and semiconductor.

On a more subtle note, we have also found that the wave-function for an electron in a crystalline material is of the form

$$\psi(x) = \sum_n C_n e^{\left(i(k+G_n)x\right)}, \tag{1.26}$$

which is simply a linear superposition of plane waves, and an example is shown in Fig. 1.11. This wave-function has no damping, which means that an electron should be able to traverse any arbitrarily large distance in the material without a change in its wave-function. Another way of putting this is to say that *an electron should encounter no resistance whilst travelling through a crystal*! This is clearly at odds with what we observe experimentally, so the only reasonable conclusion we can draw is that real materials must not be perfect crystals. In fact, *anything* which disturbs the periodicity of a crystal will give rise to electrical resistance, by scattering the electrons. There are two dominant sources of scattering:

- Defects: a missing or substituted atom will locally distort the atomic potential
- Phonons: lattice vibrations which cause the atoms to jiggle around about their mean positions, continually distorting the atomic potential and destroying the periodicity.

The resistance due to defects is temperature-independent (known as residual resistance, and is sample-dependent), whereas that due to phonons has a very strong temperature dependence, becoming approximately linear above around 100 K. When dealing with conventional resistors, the widely-used assumption is that resistance scales linearly with temperature [5], i.e.

$$R = R_0 + \alpha \Delta T, \qquad (1.27)$$

where α is the temperature coefficient of resistance of the material. In the case where the power dissipation in the resistor is large enough to cause heating, then $\Delta T \propto$ Power $\propto I^2$ which means $R = R_0/(1 - \alpha \beta I^2)$, where β is the proportionality factor between temperature and power. This relationship, whilst empirical is extremely useful, and we will encounter it again when we are looking into transport in nanowires. For now, we will consider how resistance *should* vary with temperature, by considering the effect of phonons. First though, we will briefly introduce the concept of electron *effective mass*.

1.4. Effective Mass

When we apply either a voltage or a magnetic field to a conductor, the electrons will move due to the Lorentz force. In fact, we have already implicitly assumed that the force on an electron due to an applied electric field is given by $F = eE$. From the relationships $v = \hbar k/m$ and energy, $E = \hbar^2 k^2/2m + V$, we find that the electron velocity, $v = (1/\hbar)dE/dk$. From classical mechanics, we have that the power dissipated by the electron in moving in an electric field E at velocity *v* is given by $dE/dt = eEv$. Combining these two relationships, we find that

$$dE = \left(\frac{eE}{\hbar}\right)\frac{dE}{dk}dt \qquad (1.28)$$

and

$$\frac{dv}{dt} = \left(\frac{1}{\hbar}\right)\frac{d^2E}{dk^2}\frac{dk}{dt}. \qquad (1.29)$$

From Eq. (1.28), we have that $\dfrac{dk}{dt} = \left(\dfrac{eE}{\hbar}\right)$, which, upon substitution into Eq. (1.29) gives the following:

$$\frac{dv}{dt} = \left(\frac{eE}{\hbar^2}\right)\frac{d^2E}{dk^2}. \qquad (1.30)$$

Comparing this to Newton's second law, i.e. that Force = mass*acceleration, i.e. $F = mdv/dt$, we can re-write the above equation as

$$F = m\frac{dv}{dt} = eE \Longrightarrow m = \frac{eE}{\frac{dv}{dt}}$$

$$\Longrightarrow m^* = \hbar^2\left(\frac{d^2E}{dk^2}\right)^{-1}. \qquad (1.31)$$

The star on the *m* denotes the fact that this is an *effective* mass, i.e. it is the mass that an electron appears to have within a material, be it a conductor or a semiconductor. The effective mass within conductors is generally approximately equal to the free-electron mass, but it can be significantly different inside a semiconductor. Within the free-electron approximation, the *E–k* dispersion relationship is quadratic, so the effective mass is a constant. However, as we have already seen in the nearly-free-electron approximation, most materials have a more complex band structure, so the concept of effective mass is less meaningful.

However, in the vicinity of band maxima and minima, the bands are approximately quadratic, so an effective mass may be assigned. As the atomic spacing within many crystals is direction-dependant, so is the band structure and therefore so is the effective mass, so we usually have an effective mass tensor. Thankfully, in many cases, the average value of effective mass over all directions can be used. Effective mass can be measured by a variety of means, but the most common are cyclotron resonance and angle-resolved photoemission (which, incidentally is how band structure is measured for materials today).

By considering band structure and values of the effective mass of electrons in different semiconductor materials, we can begin to understand why certain materials are so widely used by the semiconductor industry. In Table 1.2, we have written the effective masses of electrons and holes in Si, Ge, and GaAs.

Table 1.2. Effective carrier masses in semiconductor materials.

Material	Electron effective mass as a proportion of free-electron mass	Hole effective mass as a proportion of free-electron mass
Si	0.36	0.81
GaAs	0.067	0.45
Ge	0.55	0.37

To appreciate the significance of these values, remember from earlier that the maximum velocity an electron reaches within a conductor under the application of an electric field E is

$$v = \frac{e\tau E}{m} = \mu E, \qquad (1.32)$$

where μ is called the *mobility*. The higher the mobility, the faster the electrons travel and ultimately the faster the device/circuit can operate. Clearly, the mobility is inversely proportional to the electron's effective mass, so high-speed devices (the most pervasive example being mobile phones and networks) tend to use GaAs and alloys of GaAs rather than

Si. Also, GaAs is a direct-gap semiconductor unlike Si, so it has applications in lasers and photodetectors. This is the main reason why GaAs and GaAlAs are used in quantum well, wire and dot devices, which we will analyse in the next chapter. Si also has applications in photonic devices, in the form of strained silicon and polysilicon (abundance of surface states lower the band-gap sufficiently to render it effectively direct-gap).

An important question then is — why is silicon still the material of choice in most semiconductor devices and in microprocessors where speed is so crucial. The answer, not surprisingly comes down to economics — Si is one of the most abundant materials in the earth's crust (27%), it is cheap and easy to process (GaAs and its derivatives are extremely expensive to deposit and process), and its oxide, SiO_2 is one of the best insulators known — it can easily be grown, has few defects, and grows epitaxially on Si.

1.5. The Origins of Electrical Resistance

We have just seen that the nearly-free-electron model of conduction predicts that a perfect conductor should have no resistance, and that the resistance we observe is due to scattering. We will explore the quantum nature of scattering from defects in the next chapter, but first we will investigate scattering from phonons, which is the dominant source of electrical resistance. In Fig. 1.14, we have summarised the main point: an electron travels through the crystal for a certain distance

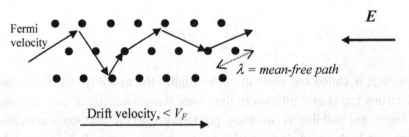

Fig. 1.14. Electron–phonon scattering. The dots represent atoms. On average, an electron will travel a distance λ before being scattered.

(the mean free path, on average) before being scattered from either a defect or a phonon. *After the scattering event, both the electron and phonon have their direction and momentum changed.* Between collisions, the electron travels at the Fermi velocity, and the transport is said to be "*Ballistic*".

To investigate the effect of phonons, we must first understand some basic points about them. We will start with the fact that the atoms in a crystal are bound together by a potential of the approximate form:

$$U(r) = 4\varepsilon \left[\left(\frac{\sigma}{r} \right)^{12} - \left(\frac{\sigma}{r} \right)^{6} \right]. \tag{1.33}$$

This is known as the *Lennard-Jones 6-12 potential* [6] for inert gases.

This potential has a minimum at $r = \sigma$, which is the equilibrium bond-length, with values between 2 and 3 Å for most materials.

The force which binds atoms together in a crystal is $-dU/dr$, which is plotted in Fig. 1.15.

For this analysis, we are just considering a 1D crystal, and are assuming the atoms are all identical.

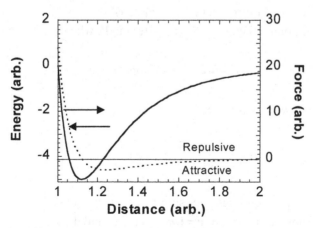

Fig. 1.15. Form of inter-atomic energy and forces. Around the equilibrium position and in contact ($F = 0$), the force depends linearly on distance *for small displacements*.

For small displacements about their equilibrium positions, however, this force is approximately linear, with the result that we can consider the atoms as being held together by springs. To understand the nature of phonons then, we must combine Schrödinger's equation with Hooke's law, i.e. we have an atom of mass m suspended from a spring of stiffness κ, which has $x = 0$ as its equilibrium position, as is illustrated in Fig. 1.16. The force on the particle, $F = -\kappa x$.

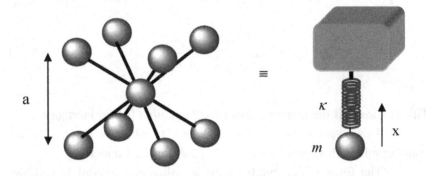

Fig. 1.16. Quantum simple harmonic oscillator. Atoms in a crystal (left) can be considered to be held together by springs (right).

According to classical mechanics, this atom on a spring will oscillate at the natural frequency $\omega_c = (\kappa/m)^{1/2}$.

The potential energy of the system is V, where

$$V = 1/2\kappa x^2 = 1/2 m\omega_c^2 x^2. \tag{1.34}$$

Schrödinger's equation for this system is

$$-\frac{\hbar^2}{2m}\frac{d^2\psi}{dx^2} + \frac{1}{2}m\omega_c^2 x^2\psi = E\psi. \tag{1.35}$$

This problem can be simplified if we employ a change of variables from x to y, where $y = (m\omega_c/\hbar)^{1/2}x$, and define $\alpha = 2E/\hbar\omega_c$.
Schrödinger's equation is now:

$$\frac{d^2\psi}{dy^2} + \left(\alpha - y^2\right)\psi = 0. \tag{1.36}$$

The solution of which is known to be $\psi(y) = F(y)\exp(-y^2/2)$ (*To see where this comes from, look at the asymptotic solution when* $y \gg a$, *which is of the form* $y = y^n \exp(-y^2/2)$.)

Substituting this form of $\psi(y)$ into the above equation, we find

$$F'' - 2yF' + (\alpha - 1)F = 0. \tag{1.37}$$

We should now assume a power series solution for $F(y)$;

$$F = \sum_{p=0}^{\infty} a_p y^p. \tag{1.38}$$

From which it can be seen that:

$$F' = \sum_{p=0}^{\infty} p a_p y^{p-1} \quad \text{and} \quad F'' = \sum_{p=0}^{\infty} p(p-1) a_p y^{p-2}. \tag{1.39}$$

An important point here is that y can never have a negative power, as that would lead to an infinity at $y = 0$ which would be unphysical (ψ must always be finite, as $|\psi(x)|^2$ represents the probability of the particle being located at position x). The first two terms of F'' therefore must equal 0, so we must put $p = p + 2$ in the above expression for F''.

Substituting for F, F' and F'' into Schrödinger's equation leads to the following:

$$\sum_{p=0}^{\infty} \left[(p+2)(p+1)a_{p+2} - (2p+1-\alpha)a_p\right] y^p = 0. \tag{1.40}$$

For a non-trivial solution, the coefficient of each power of y must vanish, leading to the following recursion relationship:

$$\frac{a_{p+2}}{a_p} = \frac{(2p+1-\alpha)}{\left[(p+1)(p+2)\right]}. \tag{1.41}$$

However, the resulting power series tends to infinity with increasing y (the limit of a_{p+2}/a_p tends to $1/p$, the sum of which is infinity), so we must truncate the power series.

The solution can actually be re-written as two power series, each containing all even or odd powers of y. Using the recursion relation, all coefficients can be expressed in terms of either a_0 or a_1. Then we need to choose some value for p, say, n, such that $2p+1-\alpha = 0$. That power series will end there, and we need to neglect the other power series. Both of these conditions lead to the following:

$$\alpha = 2n+1 \text{ for } n = 0, 1, 2...,$$
$$a_1 = 0 \text{ for } n \text{ even}, \ a_0 = 0 \text{ for } n \text{ odd}.$$

From our definition of $\alpha = 2E/\hbar\omega_c$ we have for the energy eigenstates of the quantum simple harmonic oscillator:

$$E_n = (n+1/2)\hbar\omega_c. \tag{1.42}$$

That is, discrete, equally spaced energy levels, with a ground state or *zero-point energy* of $(1/2)\hbar\omega$. Each energy level corresponds to a phonon mode.

What is the consequence of the zero-point Energy? It means that according to quantum mechanics, a harmonic oscillator can never be completely at rest, because then we would know it is momentum (zero) and position precisely, which goes against Heisenberg's Uncertainty principle. It means that even at absolute zero, the atoms in a material will still be jiggling around by a very small amount.

Normalising (i.e. using the condition that $\int_{-\infty}^{\infty} |\psi(x)|^2 \, dx = 1$), we find the first three wave functions are

$$\left.\begin{array}{l} \psi_0(x) = (m\omega_c/\pi\hbar)1/4\exp(-m\omega_c x^2/2\hbar), \\[2mm] \psi_1(x) = (4\pi)x/4(m\omega_c/\pi\hbar)3/4\exp(-m\omega_c x^2/2\hbar), \\[2mm] \psi_2(x) = (m\omega_c/4\pi\hbar)1/4(2m\omega_c/\hbar)x^2 - 1)\exp(-m\omega_c x^2/\hbar). \end{array}\right\} \tag{1.43}$$

In the ground state, the most probable position is in the centre, whereas for higher levels, the probability oscillates quickly, as shown in Fig. 1.17. Comparing to the conventional (classical) simple harmonic oscillator: the most probable position is always at the extremes (the velocity is lowest there). This is just another example of a difference between classical and quantum descriptions of the same system: they often disagree. However, when looking at highly-excited states of the quantum harmonic oscillator, the envelope of the probability distribution approaches that expected classically. This is quite a common occurrence: the quantum mechanical description of systems tends to converge towards the classical one for highly excited states.

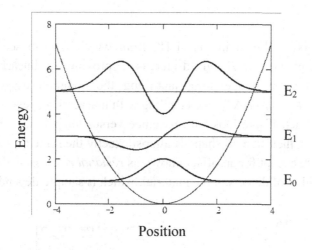

Fig. 1.17. First three levels of the quantum simple harmonic oscillator, with associated wave-functions. The oscillator potential is shown as a dotted curve.

How many phonons are present in a lattice? That is, if we assume a constant electron–phonon coupling strength and probability, then the temperature dependence of this contribution to resistance can be calculated by knowing the number of phonons.

The probability of occupation of the nth energy level due to thermal excitation is $e^{(-E_n/k_BT)}$ (*cf. Boltzmann factor*).

Therefore, we can write the average internal energy of the crystal as:

$$\langle E \rangle = \frac{\sum_{n=0}^{\infty} E_n e^{-E_n/k_B T}}{\sum_{n=0}^{\infty} e^{-E_n/k_B T}}. \tag{1.44}$$

Given that the spacing between energy levels is $\hbar\omega_c$ and the energy per oscillator is $\hbar\omega_c$ we can write the phonon occupation number $\langle n \rangle$ as $\langle E \rangle/\hbar\omega_c$, i.e.

$$\langle n \rangle = \frac{1}{e^{\hbar\omega/k_B T} - 1}. \tag{1.45}$$

This is plotted in Fig. 1.18, from which we can see that the number of phonons does indeed increase approximately linearly around and above room temperature, justifying the common usage of the expression $R = R_0 + \alpha\Delta T$ when dealing with macroscopic resistors.

In fact, if one looks at resistance versus temperature curves for most metals, their overall shape is almost exactly the same as that shown in Fig. 1.18, except for an offset known as *residual resistance*, dependent on the number of defects in the material, which is sample-dependent.

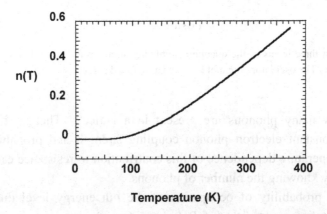

Fig. 1.18. Calculated phonon occupation number as a function of temperature.

In summary, we have seen that electrical resistance is due to the non-ideal nature of real materials, and that we can gain a basic understanding of electronic transport in conductors using quantum mechanics. A key feature of the quantum universe is the idea of quantisation, or "things being discrete", i.e. coming in small packets.

1.6. Size Effects on Electrical Resistance

The question which we are now able to address, and which is a key topic in this text, is *"how does the resistance change as we make a conductor smaller"*? What we have seen already, particularly through Fig. 1.5, is that quantum size effects tend to go unnoticed for dimensions above around 10 nm. Assuming that Ohm's law were to hold true for all dimensions, the resistance of a gold wire as a function of its diameter *should* vary as shown in Fig. 1.19.

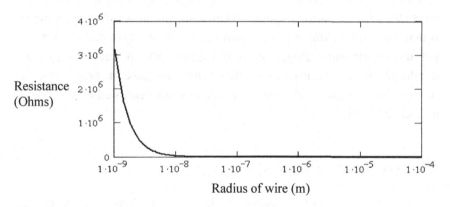

Fig. 1.19. Calculated resistance of a gold wire of length 1 mm, as its radius varies from 1 nm to 0.1 mm, using $R = \rho l / A$.

Indeed, this holds true for dimensions down to around 50–60 nm, but then begins to change dramatically, initially due to mesoscopic effects, and finally due to quantum effects. Resistance versus size in the range 1–100 nm is illustrated in Fig. 1.20.

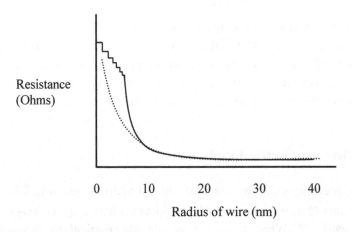

Fig. 1.20. Dependence of resistance of a wire as its radius varies from 1 nm to 40 nm. The dotted curve is that expected from the expression $R = \rho l / A$.

So far, we have restricted ourselves to looking at bulk materials and wires. In real devices of course it is the entity to which the wires are connected that is important. In the vast majority of cases, this entity is a transistor in some form or other. What is so special about the nanoscale is that a wire is no longer just a passive component which carries current, but many intriguing things start to happen, and it becomes just as important as the "transistor". We will now take a brief look at semiconductor transistors, and later chapters are dedicated to wires and molecular transistors.

1.7. Overview of Transistors

Since its humble beginnings in the early 20th century in the guise of the vacuum tube, the transistor has become the single most important electronic component on the planet. It is at the heart of operation of most electrical devices from energy-saving light bulbs to computers, mobile phones and televisions. In the first 40 years of its development, very little changed in the appearance and operation of the vacuum tube. It was only in the 1940s that all changed with the advent of the solid-state transistor by Bardeen, Brattain and Schokley at Bell

Labs. [7]. The first transistor was several cm across, whereas the transistors now in high-end computer microprocessors are around 1 μm across, the active region being less than 100 nm across.

Transistors broadly fall into two classes: bipolar junction transistors (BJTs) and field-effect transistors (FETs). The former are typically used in high-current, high-power applications, whereas the latter are used in low-current precision devices, most notably microprocessors. We will therefore concentrate on FETs as they are by far the most technologically relevant. All transistors are three-terminal devices where the current between two of the terminals is controlled by a small current (BJT) or voltage (FET) into or at the third, controlling terminal. In a BJT, the terminals are called the Base (controlling terminal), Collector and Emitter, whereas in the FET, the current flowing between the Drain and Source is controlled by the voltage at the Gate. The basic principle of operation of a FET is illustrated in Fig. 1.21. The current flows in a *channel* between the drain and source, and the length of this channel is commonly termed the *gate length*. The gate region is adjacent to this channel and an electrode is patterned on top of this and is electrically isolated from it by a thin layer of insulator, the *gate oxide*.

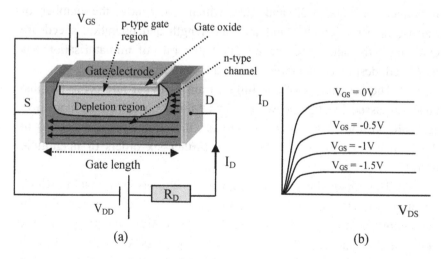

(a) (b)

Fig. 1.21. (a) Schematic of *n*-channel FET (*S* and *D* are the source and drain, respectively, *G* refers to the gate); (b) current–voltage characteristics of FET.

The width of the channel is controlled via the gate voltage; if the gate–source junction is reverse-biased then as V_{GS} increases by an amount ΔV_{GS}, the depletion region (shown in Fig. 1.21) will spread out into the channel, increasing the channel's resistance and hence decreasing the drain current, I_D. The resistor R_D is chosen such that the resulting change in voltage across it, i.e. $\Delta I_D R_D > \Delta V_{GS}$. In this way, the circuit has *amplification* — a small input signal produces a large output signal. The entire device is typically made from silicon which is doped in one of two possible configurations: the channel is *n*-type and the gate is *p*-type or vice-versa. The gate oxide is SiO_2, which has excellent electrical characteristics. We have illustrated a device with an *n*-type channel in Fig. 1.21(a), as this is the most commonly found one. In Fig. 1.21(b), we show the typically observed electrical characteristics of such an FET.

The speed of operation of a computer depends inversely on the size of the transistors within the microprocessor, and on the number of transistors. Since the late 1960s, the gate length has decreased by several orders of magnitude, and the number of transistors per microchip has increased by a factor of 2 around every 18 months. This trend is famously known as "Moore's Law", after Gordon Moore, the co-founder of Intel who first pronounced it back in the late 1960s [8]. This is illustrated in Figs. 1.22 and 1.23 which show how the number of transistors per microchip and the gate length and operation speed are evolving. In fact, since around 2000, the rate of miniaturisation has increased, despite early predictions to the contrary!

The benchmark of one billion transistors per microchip has just been surpassed in 2005/06 with Intel's Montecito chip, weighing in at 1.72 billion transistors. For comparison, the number of connections in the human brain is around 10^{15} — a further six orders of magnitude more!

The dimensions of transistors are approaching that at which quantum effects will become noticeable. In fact, this is already happening, as the gate oxide is typically 0.8 nm thick, so current can leak between the gate and the channel, via a process known as *quantum tunneling*, which we will look at in more detail in the next chapter. The gate length is below 50 nm made using 90 nm process technology, 35 nm using 65 nm technology, and < 25 nm using 45 nm technology, which is due to come on-line in 2007 [9]. Prototype transistors with gate lengths in the range 10–15 nm have already been produced and have been shown to have excellent performance. IBM have fabricated a proof-of principle transistor with a gate length of 6 nm [10].

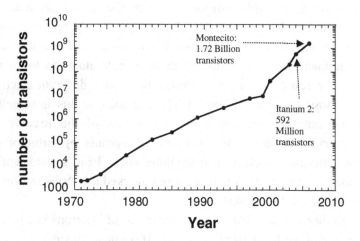

Fig. 1.22. Chronological development of the number of transistors per microchip (Moore's law).

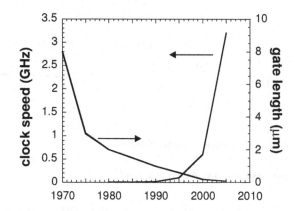

Fig. 1.23. Chronological development of the gate length and speed of transistors.

1.8. Surface Effects

The famous physicist Wolfgang Pauli once said that *"God made solids but surfaces were the work of the devil"*. This statement was prompted by the fact that the properties of surfaces are very different to those of the bulk. The reason for this is of course that while atoms in the bulk are surrounded by other atoms on all sides, i.e. they are fully co-ordinated and in stable positions, those on the surface are not. This broken symmetry leads to surface atoms being highly reactive, and in response to the lack of co-ordination, on many surfaces the atoms re-arrange their positions relative to those of bulk atoms to lower their energy. The resulting surface reconstructions have different electronic states to those found in the bulk [11]. Fortunately, this is something which we can take advantage of, and is one of the reasons why nanostructures have drawn so much interest, because by virtue of their size, they are more surface than anything else. Pauli's statement has been superceded by E. W. Plummer who remarked that *"surfaces are the playground of solid state physics"*.

Earlier, we saw that the wave-functions of electrons in a periodic potential are of the form $\psi(x) = e^{ikx}u(x)$. If we now consider the case of a 3D semi-infinite solid, as shown in Fig. 1.24, the wave-functions are:

$$\psi_k(r) = e^{ik \cdot r}u_k(r). \tag{1.46}$$

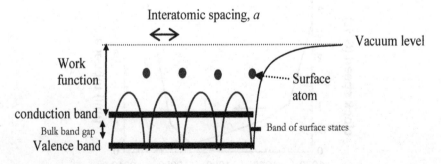

Fig. 1.24. Lattice potential near a surface. The break in symmetry caused by the surface induces the formation of surface states known as *Tamm* states.

Now, we have implicitly assumed in our previous analysis that the k vector is real. If it were to be imaginary, then $\psi_k(r)$ could diverge to infinity within the crystal. One possibility is to make the component of k parallel to the surface ($k_\parallel$) real whilst making $k_\perp$ imaginary, i.e.

$$\psi_k^s(r) = e^{ik_\parallel \cdot r_\parallel} u_{k_\parallel}(r_\parallel) e^{-k_\perp \cdot r_\perp}. \tag{1.47}$$

This results in wave-functions which decay exponentially away from the surface both into the bulk and into free space (these wave-functions are bound to the surface and are called *surface states*). One therefore needs to find wave-functions within the bulk which show this property of growing exponentially towards the surface, as is illustrated in Fig. 1.25.

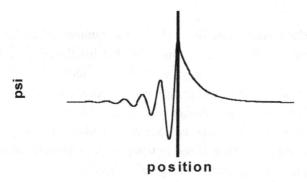

Fig. 1.25. Surface state wave-function. The vertical line indicates the position of the surface. The envelope of the wave-function (psi) decays exponentially away from the surface.

As one might expect, for any given value of $k_\parallel$ there is a finite spectrum of allowed $k_\perp$ values. Thus, the spectrum of surface state energy levels is discrete. One point to note is that surface states are non-degenerate with bulk states, as then it would be possible for an electron in a surface state to couple to the bulk and there would no longer be any confinement of electrons on the surface. Therefore, *surface state energy levels exist within the bulk band-gaps.* Surface states broadly fall into two classes, *Shockley* and *Tamm* states. Basically, Tamm states are due to the broken periodicity of a crystal due to the presence of a surface, whereas Shockley states arise due to the re-arrangement of surface atoms. The origin of both types of state are illustrated in Figs. 1.24 and 1.26. Therefore, whilst both tend to co-exist, the relative amount of the two types of state depends on the nature of the surface atomic arrangement.

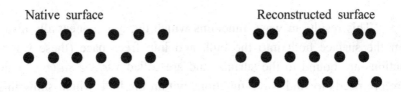

Fig. 1.26.   Re-arrangement of surface atoms (*surface reconstruction*) induces the formation of *Shockley* surface states.

Surface states can be probed in a number of different ways which are beyond the scope of this book, but briefly they are based on performing energy spectroscopy on electrons which have been ejected from the surface, either using photons — XPS (X-ray photoelectron spectroscopy), electrons (Auger spectroscopy), or ions.   Surface reconstructions may be observed locally in real-space using scanning-probe microscopy, as we will see in Chapter 4, or globally in reciprocal space using low energy electron diffraction (leed).

References for Chapter 1

1.   G. Ohm, *Die galvanische Kette: mathematisch bearbeitet* (*The Galvanic Circuit Investigated Mathematically*) (Riemann, Berlin, 1827).
2.   N. W. Ashcroft and N. D. Mermin, *Solid State Physics* (Saunders College Publishing, Philadelphia, 1976).
3.   S. M. McMurry, *Quantum Mechanics* (Addison-Wesley, 1993).
4.   F. Bloch, *Z. Physik* **52**, 555–600 (1928).
5.   P. W. Horowitz and W. Hill, *The Art of Electronics* (Cambridge University Press, 1980).

6. J. E. Lennard-Jones, *Proc. Phys. Soc.* **43**, 461–482 (1931).
7. The junction transistor, US Patent #02569347.
8. G. E. Moore, *Electronics* **38** (1965).
9. Source: Intel website.
10. Source: IBM website.
11. An excellent introduction to surface science is A. Zangwil, *Physics at Surfaces* (Cambridge University Press, Cambridge, 1988).

Chapter 2

Quantum Current Flow

In 1959 when Richard Feynman delivered his prophetic talk "there's plenty of room at the bottom" [1] he prompted us to consider the possibility of building devices *from the bottom up*, i.e. manipulating single atoms into functional structures capable of performing specific tasks. That vision has inspired generations of scientists, and has certainly contributed to the impetus driving us to make ever smaller devices. The pinnacle of this sort of work is already within our grasp, in the form of single-molecule devices and single-electron devices, namely transistors. Figure 2.1 shows the first solid-state transistor for comparison with a contemporary one, illustrating that over the past fifty years, the transistor has decreased in size by nearly six orders of magnitude.

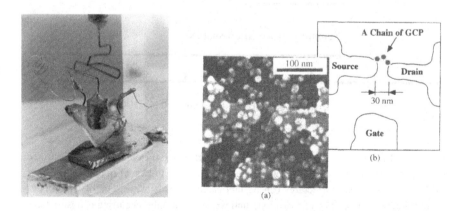

Fig. 2.1. The first solid-state transistor [2] and a single-electron transistor [3].

As the length-scale of interest within these devices is of the order 1–10 nm (see Fig. 2.2), we find ourselves in the situation whereby we need new tools to fabricate and characterise them. Conventional Scanning or Transmission Electron Microscopy (S(T)EM) has been used for this type of analysis in the past. One can even utilise voltage contrast to observe the distribution of electric potentials around a working circuit. The inevitable presence of metals with oxides however typically limits the resolution of this technique to several tens of nm due to charging effects. To obtain a better spatial resolution requires a conductive overcoat, which in turn renders voltage contrast impractical. We will see shortly that this particular obstacle has been spectacularly overcome with the use of scanning probe microscopes, which we will consider in Chapter 4.

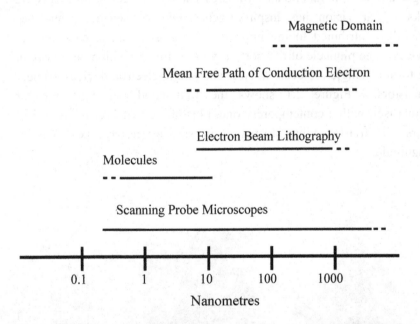

Fig. 2.2. Physical length-scales relevant to nanoscience and nanotechnology. Electronic mean free paths range from a few nm at room temperature to tens of microns or more at low temperature. Nanostructures can be fabricated in many ways, one of which is e-beam lithography (top-down fabrication), and we have naturally occurring nanostructures already — molecules. We can probe the nanoworld using scanning probe microscopes.

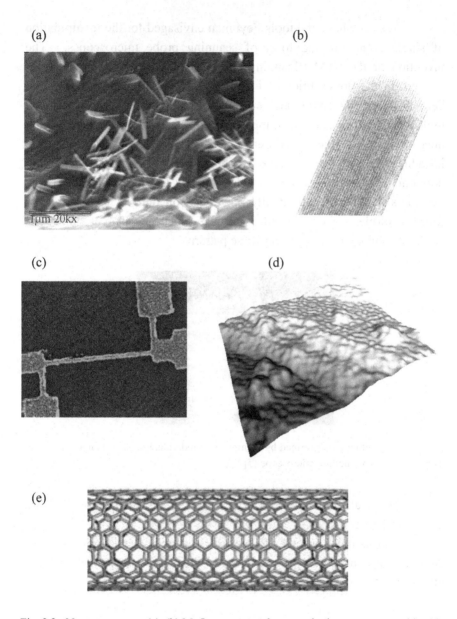

Fig. 2.3. Nanostructures. (a), (b) MoOx nanocrystals grown by bottom-up assembly, (c) Au nanowire fabricated by electron-beam lithography, (d) Lander molecules near atomic steps on Si(111) surface imaged by STM, (e) schematic of a carbon nanotube.

We now have the tools Feynman envisaged for the manipulation of single atoms, in the form of scanning probe microscopes. The invention of the STM (Scanning Tunnelling Microscope) in 1982 [4] opened up the possibilities of interactive science on the nm scale and below. Not only could one image and manipulate single atoms and molecules, but one could also measure their electrical properties through measurements of the local density of states, a quantity which we introduced in Chapter 1. Figure 2.4 shows the "quantum coral" showing how one can tailor electronic states using this instrument. In this case, Fe atoms have been individually positioned on the Cu(111) surface to form a circle. Consequently, surface electronic waves/states are confined, and we get a standing wave pattern.

Fig. 2.4. A quantum coral, formed by manipulating individual Fe atoms on a Cu surface using a scanning tunnelling microscope [5].

We shall return to look at scanning probe microscopy in more detail in Chapter 4.

At this point, in order to see the relevance of developing smaller devices and consequently tools to study them, we briefly consider the effects incurred by shrinking devices, i.e. why bother?

2.1. Why Shrink Devices?

The argument for shrinking devices is based on the fact that electrons travel around a circuit at drift velocities of the order 10^5 m.s^{-1}.

As the typical distance from one transistor to its nearest neighbour is a few μm (see Fig. 2.5), the transit time is of the order 0.1 ns, meaning the current limit to processor speed is of order 10 GHz. To speed devices up, either the carrier mobility must increase (by for instance reducing the amount of defects or lowering the temperature), and/or the transit distances must decrease.

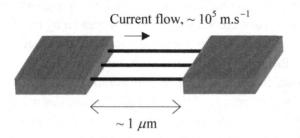

Current flow, $\sim 10^5$ m.s^{-1}

$\sim 1 \ \mu$m

Fig. 2.5. Two devices separated by 1 μm can communicate with each other at a rate of ~ 10 GHz. Higher speeds can be achieved by increasing the mobility of the charge carriers and decreasing the device size and separation.

In a conventional transistor, the effective current pathway (i.e. the gate length) is of the order 180 nm. In terms of the characteristic length scales involved in conduction (e.g. mean free path, coherence length, screening length, etc.), that is reasonably large, so the overall motion of the conduction electrons can be described as essentially diffuse, as random scattering processes remove any phase coherence. This type of transport is well described using the Boltzmann equation [6], a key ingredient being the relaxation time, which determines the mean free path (mfp). The inelastic mfp is typically a few 10 nm at room temperature in a good conductor, and the phase coherence length is generally larger (a single scattering event usually only partly randomises the phase, hence there is generally some degree of phase coherence for several mfps). Clearly, as devices get smaller, the transport will no longer be strictly diffuse, but is *ballistic*, and the Boltzmann equation will be inapplicable. Individual scattering events will now be important rather than an average over many events. Attempts to model this type of transport are based on the Landauer–Büttiker formalism [7], which considers conduction in terms of quantum-mechanical transmission probabilities for electrons to traverse a current pathway. At this length-scale, the phase coherence of the conduction electrons causing quantum interference effects and conductance fluctuations, and capacitative effects causing Coulomb Blockade (which we shall discuss later) all combine to give rise to complex current–voltage characteristics.

If two transistors could now be located closer together than the mfp, the processor speed could be potentially increased by two orders of magnitude, as the carriers are travelling at the Fermi velocity (i.e. ballistically), of the order $1.4*10^6$ m.s^{-1}. As single scattering events are now so important, the probes used to perform electrical measurements at this scale will strongly influence the measurement outcome. Consequently the probes should be as small as possible, and certainly smaller than the device of interest. Interpretation of measurements is aided by use of the Büttiker formula, which allows us to describe transport properties in terms of measurable currents and voltages. We shall return to this later with specific examples.

2.2. Point Contacts: From Mesoscopic to Atomic

Let us now consider what happens when we have a point contact between two metals, as shown in Fig. 2.6.

On the basis of continuum models, we would expect that if we were to shrink the tip towards an infinitely small point, the resistance would increase as $1/r^2$, r being the radius of the contact. However, we

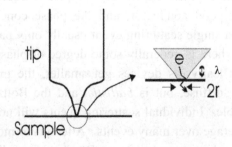

Fig. 2.6. A point contact of radius r. The electrons (e) have a mean free path of λ.

find that when the dimensions of the contact are comparable to the mean free path, there are departures from this behaviour.

When the contact radius is large compared to the mfp, the resistance is given by:

$$R_S = \frac{\rho}{2r}. \tag{2.1}$$

This is known as the *"spreading resistance"* [8]. As we shrink a conductor to well below the mfp, the resistance also departs from the expected value. An approximate formula is given by Sharvin [9] as

$$R = \frac{4\rho\lambda}{3\pi r^2}, \tag{2.2}$$

where ρ is the resistivity and λ is the electronic mfp. This is valid when $r \ll \lambda$. Both are plotted in Fig. 2.7. However, this continuum approach breaks down once the contact area approaches atomic dimensions, as we saw at the end of Chapter 1. This is due to the bound energy states of the electrons confined in the constriction having discrete energy levels. We saw from before that the size at which the spacing between these levels is comparable to $k_B T$ is around 10 nm.

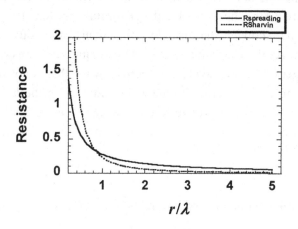

Fig. 2.7. Resistance of a point contact by Sharvin's formula, and spreading resistance.

The form of the potential which confines electrons in a constriction is approximately parabolic (just as it was for the quantum harmonic oscillator) so the bound energy states, or *modes* are equally spaced. These modes can only exist when the constriction size is an integer multiple of half Fermi wavelengths (λ_f) across. Therefore, the number of modes *per dimension* is given by

$$M = Int\left[\frac{2W}{\lambda_f}\right],\qquad (2.3)$$

where W is the width of the contact. As a consequence, the conductance is a stepwise function of contact size rather than a monotonic one. As we will see later, the Landauer formula [7] expresses the conductance as

$$G = \frac{2e^2}{h}\sum_{i=1}^{M} T_i,\qquad (2.4)$$

where T_i is the quantum-mechanical probability for an electron to be transmitted through the ith channel in the constriction. There are numerous experimental verifications of this formula in the literature [10,11]. Figure 2.8(a) shows the Landauer formula plotted alongside the Sharvin formula, showing the discrete rather than continuous nature of the resistance. As the electronic energy states require the constriction to be a half integer of Fermi wavelengths across, when that is not satisfied, considerable forces are exerted by the electronic gas on the lattice, and the constriction will tend to resize to compensate. Figure 2.8(b) shows the result of a conductance and force measurement for the necking of a Au tip [after Ref. 12].

2.3. Conductance from Transmission

We have already seen in Chapter 1 that by using quantum mechanics, we can explain a range of observable properties of materials.

We were also able to derive Ohm's law. How then, can we formally use quantum mechanics to describe current flow in a system? At the heart of this is the concept of *scattering*. We do not need quantum mechanics to tell us that in the absence of scattering, electrons will flow unhindered from one place to another. We do need quantum mechanics however to understand *how* scattering occurs and the effect it has. Whenever a particle encounters any sort of discontinuity in its environment, it will be scattered from its original path, and there is a finite chance that it will be completely reflected along its incident path. A measure of the amount of scattering in a system is given the lofty name of *transmission probability*, *T*. If we apply a voltage between two points of a circuit, as shown below in Fig. 2.9, a current will flow, and this current depends on the voltage difference as well as on *T*.

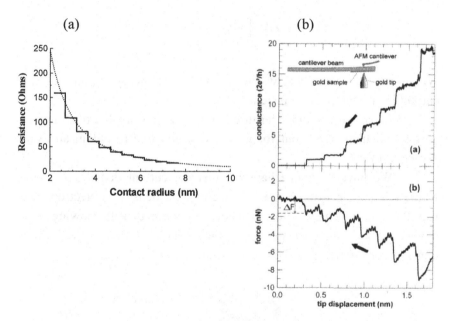

Fig. 2.8. (a) Resistance of a point contact by Sharvin's formula together with that predicted by the Landauer formula; (b) conductance and force versus tip displacement for a gold neck formed by pushing a gold tip into a gold surface which is itself on a cantilever (from Ref. 12). The conductance changes by integer multiples of the conductance quantum due to increases in the contact area.

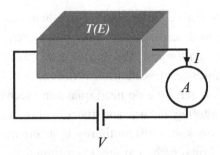

Fig. 2.9. Current flow through an arbitrary system characterised by a transmission probability, *T*.

The transmission probability is generally a function of the electron energy, so the current will be of the form:

$$I \propto \int_{eV}^{0} T(E)dE. \qquad (2.5)$$

We say that the current flows in "channels" of energy, *E*, so the total current is the sum over all the channels.

Later, we will consider through the use of a number of examples how to calculate *T(E)*, but for now, we would like to obtain an exact expression for *I*.

We have seen in Chapter 1 that the current density, $J = env$, where *e* is the electron charge, *n* is the number density of electrons and *v* is the drift velocity. Now consider a current density flowing with momentum k_x in the positive *x* direction J_{left} incident on an object with a transmission coefficient, *T(E)*, as illustrated in Fig. 2.10.

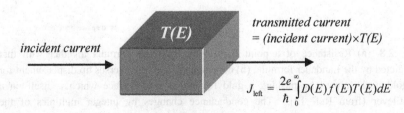

Fig. 2.10. Current through a quantum system.

We still have $J_{\text{left}} = env_x$, but the difference now is in what we mean by the quantity, n. If we consider electrons within dk_x of k_x it will be given by the density of states in k-space, $\mathcal{D}(k)$ times dk, and we will also need to include the occupation probability of the states, $f(k)$, which in most cases will be the Fermi function. In other words, $n = \mathcal{D}(k)f(k)dk$. From this, we can write the incident current density within a narrow momentum range dk as,

$$J_{\text{left}} = ev_x D(k) f(k) dk. \tag{2.6}$$

(*The density of states in a 3D k-space is just $2/(2\pi)^3$, where the factor of 2 comes from the two possible spin states. For now, we will only incorporate the factor of 2, but otherwise leave $\mathcal{D}(k)$ as an arbitrary function*).

We have already ascertained that the 3D density of states is given by:

$$\mathcal{D}(E)_{3D} = \frac{dN}{dE} = \frac{E^{1/2}}{2\pi^2}\left(\frac{2m}{\hbar^2}\right)^{3/2}.$$

We can simplify Eq. (2.6) using the relationship that momentum = mv = $\hbar k$. The *net* current which flows through the object from left to right is then J_{left} times the transmission probability, $T(k)$, integrated over all possible k values (must be positive, as a negative k would represent electrons flowing from right to left) i.e.

$$J_{\text{left}} = \frac{2e\hbar}{m} \int_0^\infty D(k) f(k) T(k) k\, dk. \tag{2.7}$$

As the quantity we can physically control in any experiment is the voltage, and hence the electron energy, we should convert our variable to energy rather than momentum (k).

We have that $E = \dfrac{\hbar^2 k^2}{2m}$ from which it follows that $dE = \dfrac{\hbar^2 k}{m} dk$.

Current at the Nanoscale

We can substitute this into the above expression to obtain

$$J_{\text{left}} = \frac{2e}{\hbar} \int_0^\infty D(E)f(E)T(E)dE. \qquad (2.8)$$

The net current density is the difference between currents flowing in both directions, i.e.

$$J_{\text{total}} = \frac{2e}{\hbar} \left(\begin{array}{l} \int_0^{\mu_l} D(E)f_l(E_l)T_l(E)dE \\[2mm] - \int_0^{\mu_r} D(E)f_r(E)T_r(E)dE \end{array} \right)$$

$$= \frac{2e}{\hbar} \left(\int_{\mu_r}^{\mu_l} D(E)\big(f_l(E) - f_r(E_r)\big)T(E)dE \right), \qquad (2.9)$$

where $f_l(E)$ and $f_r(E)$ are the Fermi functions on the left- and right-hand contacts, respectively, and E_r is the energy on the right-hand side relative to the left-hand side, which will be $E_l + eV$, where V is the applied voltage. Also, a feature of quantum systems is that the transmission probability, $T(E)$ is symmetric with respect to direction, so $T_l(E) = T_r(E)$.

The next step in our analysis is to determine the form of the density of states, $\mathcal{D}(E)$ as a function of the dimensionality of the system of interest. To illustrate the point, Fig. 2.11 shows $\mathcal{D}(E)$ versus E for 0, 1, 2 and 3 dimensions, plotted using the formulae 1.7–1.10.

Therefore, for the 1D case (current density has no meaning in 1D, so we replace J with I),

$$I^{1D}_{\text{left}} = \frac{2e}{h} \int_0^\infty f(E)T(E)dE. \qquad (2.10)$$

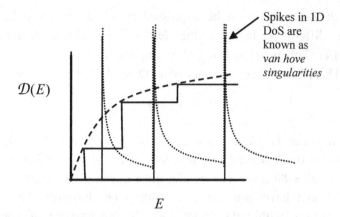

Fig. 2.11. Density of electronic states for 3D (dashed curve), 2D (solid lines), 1D (dotted curves) and 0D (solid spikes).

Similarly, currents injected from the right are of the form

$$I^{1D}{}_{right} = \frac{2e}{h} \int_0^\infty f'(E)T'(E)dE. \tag{2.11}$$

If we also assume we are at low temperature, then $f(E) = 1$ for $E < \mu_l$, 0 for $E > \mu_l$, and $f'(E) = 1$ for $E < \mu_r$, 0 for $E > \mu_r$, where $\mu_{l,r}$ are the electrochemical potentials of the left- and right-hand contacts, respectively.

As a final result, the net current flowing through the 1D quantum system at Zero Kelvin is $I_{left} - I_{right}$, or

$$I^{1D} = \frac{2e}{h} \left(\int_0^{\mu_l} T(E)dE - \int_0^{\mu_r} T(E)dE \right) = \frac{2e}{h} \left(\int_{\mu_r}^{\mu_l} T(E)dE \right). \tag{2.12}$$

If $T(E)$ does not vary significantly over the energy range of interest, then

$$I^{1D} = (2e/h)T(\mu_l - \mu_r). \tag{2.13}$$

The conductance can then be expressed as I/p.d., where p.d. is the potential difference between the left-and right-hand contacts, or $(\mu_l - \mu_r)/e$. This is another way of writing Ohm's law.

Therefore, the conductance of a 1D conductor is given by:

$$G^{1D} = (2e^2/h)T. \qquad (2.14)$$

This is known as *Landauer's formula*, and shows that in fact, electrical conductance is *quantised*, in multiples of $2e^2/h$ (which has the numerical value 80 μS). The conductance quantum is variable only in that $T(E)$ can have any value, depending on the particulars of the potential profile within the conductor. In the absence of scattering (a special situation known as ballistic conduction), $T = 1$.

This is a general result and can be applied to any transport problem. There are a number of key assumptions that we have made to get to this point which should be re-examined. First, we have assumed that there is only one conductance channel, or one electron "*mode*" within the confines of our conductor. There will generally be very many of these, as the Fermi wavelength (the wavelength of electrons at the Fermi energy) is typically around 0.5 nm, so in a conductor of width W, there will be $2W/\lambda_F$ modes. The conductance will be increased relative to the single-channel case by this factor.

Another assumption we have made is that both contact regions are electron reservoirs with an infinite number of electronic states, and the current is *probing* the wire joining those contacts. This is of course not the case. For tunnelling, the region between the contacts is a vacuum rather than a conductor, and the density of states of the vacuum is taken as a constant which we may ignore, as we will see later with the scanning tunnelling microscope (STM). In that case, the current becomes

$$J_{\text{total}} = \frac{2e}{\hbar}\left(\int_{\mu_r}^{\mu_l} D_l(E)D_r(E)\big(f_l(E) - f_r(E_r)\big)T(E)dE \right). \qquad (2.15)$$

The difference now being that there is no wire joining the contacts, and the current is actually due to the contacts probing each other.

2.4. Calculation of Transmission Probability and Current Flow in Quantum Systems

2.4.1. *Introduction to the concept of transmission probability*

We stated earlier that whenever a quantum particle encounters a discontinuity in its surroundings, it will be scattered. This is illustrated in Fig. 2.12.

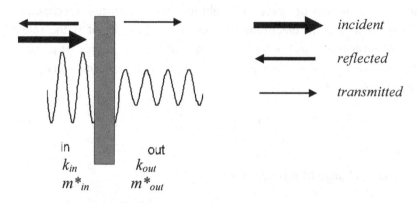

Fig. 2.12. Illustration of scattering of a quantum particle from a potential discontinuity. k_{in}, k_{out}, m^*_{in}, and m^*_{out} are the k-vectors and effective masses of the incident and transmitted electrons. The size of the arrows represent the relative amplitudes of the incident, reflected and transmitted wave-functions.

In order to investigate the amount of scattering, we must use one of the fundamental postulates of quantum mechanics, which is that *the wave-function describing a particle must be a continuous function*, so that when we have several boundaries separating different regions of space, the wave-function may have a different form in each region, but these functions *and their first spatial derivatives divided by the electron's effective mass* must match at the boundaries. To see why this is the case, $|\psi(x)|^2$ represents the probability of finding the particle at position x, which must be single-valued, which can only happen if ψ

itself is continuous. The kinetic energy operator is proportional to $(1/m^*)\partial^2\psi/\partial x^2$, so in order to always have a finite kinetic energy, $(1/m^*)\partial\psi/\partial x$ must always be continuous.

Using these two conditions (known as the BenDaniel–Duke boundary conditions [13]), we can solve for the unknowns in the wave-function, which we get from solving Schrödinger's equation.

To calculate the probability of a particle being reflected or transmitted from a given potential discontinuity, we only need to look at the relative amplitudes of the probability flux before and after passing through that region of space. Probability flux in quantum mechanics is analogous to flux in electromagnetism or gas kinetics, it is the rate of flow of probability, as opposed to the rate of flow of energy or particles. The probability of finding a particle with wave-function $\psi(r)$ in a volume v is P, which is given by

$$P = \int_v |\psi|^2 d^3r. \tag{2.16}$$

The rate of change of this probability is

$$\frac{dP}{dt} = \int_v \frac{\partial}{\partial t}|\psi|^2 d^3r = \int_v d^3r \left\{ \psi^* \frac{\partial}{\partial t}\psi + \left(\frac{\partial}{\partial t}\psi^*\right)\psi \right\}. \tag{2.17}$$

However, from Schrödinger's time-dependent equation, we know that

$$-i\hbar\frac{\partial\psi}{\partial t} = -\frac{\hbar^2}{2m^*}\nabla^2\psi + V(r)\psi. \tag{2.18}$$

And similarly for the complex conjugate of ψ. If we substitute this into the above equation, we obtain:

$$\frac{dP}{dt} = \frac{i\hbar}{2m^*}\int_v d^3r \left[\left\{\psi^*\left(\nabla^2\psi\right) - \left(\nabla^2\psi^*\right)\psi\right\} + \left\{\psi^*V\psi - V\psi^*\psi\right\} \right]. \tag{2.19}$$

The terms inside the second set of brackets in the integral cancel, so we are left with

$$\frac{dP}{dt} = \int_v \frac{\partial}{\partial t}|\psi|^2 d^3r = \frac{i\hbar}{2m^*}\int_v d^3r \nabla \bullet \{\psi^*(\nabla\psi) - (\nabla\psi^*)\psi\}. \quad (2.20)$$

Now the integrands must be equal for any volume, so we have:

$$\frac{\partial}{\partial t}|\psi|^2 = \frac{i\hbar}{2m^*}\nabla \bullet \{\psi^*(\nabla\psi) - (\nabla\psi^*)\psi\} = \frac{i\hbar}{2m^*}\nabla \bullet j. \quad (2.21)$$

This has the form of a continuity equation, where the quantity, j, is the probability flux. For plane wave wave-functions, $j = |\psi|^2 \hbar k/m^*$. Therefore, for any quantum scattering problem, the transmission probability is simply the ratio of the transmitted to the incident *flux*.

2.4.2. *Single potential step*

The first problem which we will consider is that of a 1D potential step, as shown in Fig. 2.13.

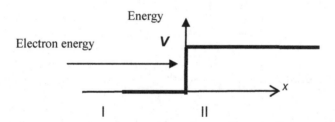

Fig. 2.13. One-dimensional potential step of height V. The potential is V when $x > 0$, and 0 when $x < 0$.

As GaAlAs has a larger band-gap than GaAs, such a device may be fabricated simply by depositing GaAlAs on top of GaAs, as shown in Fig. 2.14. The band-gap (at room temp.) of GaAs is 1.43 eV, whereas $Ga_{1-x}Al_xAs$ is $(1.424 + 1.247x)$ eV when $x < 0.45$, and $(1.9 + 0.125x + 0.143x^2)$ eV when $x > 0.45$. Also, when $x < 0.4$, the band-gap in $Ga_{1-x}Al_xAs$ is direct [14].

We are initially interested in solutions when $E < V$ (i.e. when electron has less energy than the step).

Fig. 2.14. A one-dimensional potential step may be fabricated by depositing GaAlAs on top of GaAs.

We can split the solution into two regions of space, I and II as shown in Fig. 2.13. The equations to solve are:

$$(-\hbar^2/2m\partial^2/\partial x^2)\Psi_{\mathrm{I}}(\mathbf{x}) = E\Psi_{\mathrm{I}}(\mathbf{x}). \qquad \textbf{Region I}$$

$$(2.22)$$

$$(-\hbar^2/2m\partial^2/\partial x^2 + V)\Psi_{\mathrm{II}}(\mathbf{x}) = E\Psi_{\mathrm{II}}(\mathbf{x}). \qquad \textbf{Region II}$$

The solutions are of the form:

$$\Psi_{\mathrm{I}} = A_1 e^{ik_1 x} + B_1 e^{-ik_1 x},$$

$$\Psi_{\mathrm{II}} = A_2 e^{k_2 x} + B_2 e^{-k_2 x},$$

$$(2.23)$$

where $k_1 = \dfrac{\sqrt{2mE}}{\hbar}$ and $k_2 = \dfrac{\sqrt{2m(V-E)}}{\hbar}$.

We can simplify this a little if we look at the form of Ψ_{II}, as the exponentially increasing part $(A_2 e^{k_2 x})$ is in fact unphysical. The step continues to $x = \infty$, which means Ψ_{II} approaches infinity. If we remember what $|\Psi(\mathbf{x})|^2$ represents, i.e. the probability of locating the particle at position x, then it must always be finite. Without any loss of generality then, we can set $A_2 = 0$.

To determine the reflection/transmission coefficients for this system, we need to know the relationship between the coefficients A_i and B_i, i.e. what are the relative strengths of the incident and reflected waves? This is where the boundary conditions are useful, as when we match the wave-functions and their first derivatives at the boundary $(x = 0)$ we obtain the following relationships:

$$A_1 + B_1 = B_2,$$

$$ik_1 A_1 - ik_1 B_1 = -k_2 B_2, \qquad (2.24)$$

i.e. $B_1/A_1 = -(k_2 + ik_1)/(k_2 - ik_1),$

where we have implicitly assumed that the electron effective mass is equal on both sides of the step — which will not be true in reality.

The reflection probability $= |B_1/A_1|^2 = 1$ identically so there is *zero* probability of the electron passing through this potential step. This result is unchanged when we assume different electron effective masses on both sides. The wave-functions are shown in Fig. 2.15.

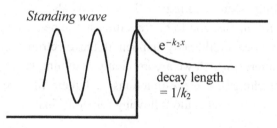

Fig. 2.15. Wave-function for an electron encountering a potential step, where the electron energy is lower than the step height.

The curious point to note is that the wave-function in region II, Ψ_{II} is not zero, but decays exponentially into region II, with a characteristic decay length of $1/k_2$ (this is the distance at which Ψ_{II} drops to $1/e$ of its maximum value). This is very much at odds with what classical, Newtonian mechanics predicts. If we invoke conservation of energy, and say that the total energy of the electron at any position is the sum of its kinetic and potential energy, then $E = K.E. + V$. In region I, $V = 0$, so $E = K.E.$ In region II, $V \neq 0$, so $K.E. = E - V$. However, in the case we have just considered, $V > E$, which means $K.E.$ is negative! How can we have a negative kinetic energy? The answer is that we cannot, so the particle gets reflected from the step, in accordance with classical mechanics. However, there is still a finite probability that the electron will be found *just inside* the step for a very short time, and this is allowed by Heisenberg's uncertainty principle [15].

How about the case when $E > V$ (i.e. the total energy of the electron is greater than the potential energy of the step)?

Well, in that case the solutions to Schrödinger's equation become:

$$\Psi_I = A_1 e^{ik_1 x} + B_1 e^{-ik_1 x}, \quad \Psi_{II} = A_2 e^{ik_2 x} + B_2 e^{-ik_2 x}, \quad (2.25)$$

where $k_1 = \dfrac{\sqrt{2mE}}{\hbar}$ and $k_2 = \dfrac{\sqrt{2m(E-V)}}{\hbar}$.

In a similar way to before, we can eliminate $B_2 e^{-ik_2 x}$, as that represents a left-travelling electron in region II, which is also unphysical — there is no reason for an electron to be travelling in the negative x-direction after the step, as that could only happen if the electron were to encounter another scatterer, which we are implicitly assuming is not the case.

Matching the wave-functions and their first derivatives at the boundary $(x = 0)$ yields the following relationships:

$$A_1 + B_2 = A_2,$$
$$ik_1 A_1 - ik_1 B_1 = ik_2 A_2, \quad (2.26)$$

i.e. $B_1/A_1 = (k_1 - k_2)/(k_1 + k_2)$ ....[*or, if we assume different electron effective masses,* = $(k_1/m_1 - k_2/m_2)/(k_1/m_1 + k_2/m_2)$].

The wave-functions and the reflection probability as a function of incident electron energy are shown below in Fig. 2.16.

What can we say about this? Well, every time a particle experiences a potential discontinuity, there is a finite probability that it will be reflected. The larger the discontinuity, the greater is the probability of reflection.

This means that every time a quantum particle experiences a potential discontinuity, there is a finite probability that it will be reflected. The larger the discontinuity, the greater is the probability of

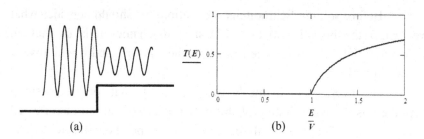

Fig. 2.16. (a) Wave-function for an electron impinging on a step of lower energy than the electron, and (b) transmission probability for an electron impinging on a potential step 2 eV high, as a function of electron energy.

reflection. This is analogous to the propagation of waves down transmission lines — whenever the wave encounters a discontinuity in impedance, it will be partially reflected. The difference there of course, is that an electromagnetic wave can be partially reflected, whereas a particle will be either completely reflected or transmitted, so what we are seeing here is the *probability* of transmission occurring for any given particle.

2.4.3. *Single potential barrier*

2.4.3.1. *Symmetric barrier: No applied voltage*

We now wish to consider the case where the step is replaced by a barrier of finite width, as shown in Fig. 2.17.

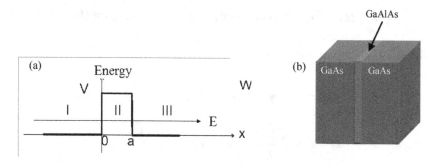

Fig. 2.17. (a) Potential barrier structure; (b) construction of barrier.

Before solving Schrödinger's equation, we should consider what we *expect* the behavior to be. Classical mechanics tells us that an incident particle will always be reflected when $E < V$, and it will always be transmitted when $E > V$.

However, quantum mechanics showed us that for the potential step, there is always a finite probability that the particle will be reflected when $E > V$. The same will be true for the potential barrier. The difference now is that if we remember the shape of the wave-function inside the step when $E < V$, it was an exponentially decaying function. We might expect then, that if the barrier is thin enough, the decaying wave-function will still have an appreciable amplitude after passing through the barrier, and that will result in a small travelling wave at the output side of the barrier. This is in fact what happens, and the probability of transmission is known to scale exponentially with the energy of the incident particle (this is due to the fact that the decay length of the wave-function in the barrier is proportional to $E^{0.5}$). This phenomenon, whereby a particle can "pass through" a barrier against the rules of classical mechanics, is known as *quantum tunnelling*, and is something we will see over and over again within this book as being an extremely important and useful phenomenon.

To gain a qualitative understanding of tunnelling, we need to solve Schrödinger's equation for the barrier, which in the three regions is:

$$(-\hbar^2/2m\partial^2/\partial x^2)\Psi_{\mathrm{I}}(x) = E\Psi_{\mathrm{I}}(x), \qquad \textbf{Region I}$$

$$(-\hbar^2/2m\partial^2/\partial x^2 + V)\Psi_{\mathrm{II}}(x) = E\Psi_{\mathrm{II}}(x), \qquad \textbf{Region II} \qquad (2.27)$$

$$(-\hbar^2/2m\partial^2/\partial x^2)\Psi_{\mathrm{III}}(x) = E\Psi_{\mathrm{III}}(x). \qquad \textbf{Region III}$$

The solutions are of the form:

$$\Psi_{\mathrm{I}} = A_1 e^{ik_1 x} + B_1 e^{-ik_1 x}, \quad \Psi_{\mathrm{II}} = A_2 e^{k_2 x} + B_2 e^{-k_2 x}, \quad \Psi_{\mathrm{III}} = A_3 e^{ik_3 x},$$

$$ (2.28) $$

$$\text{where} \qquad k_1 = k_3 = \frac{\sqrt{2mE}}{\hbar} \quad \text{and} \quad k_2 = \frac{\sqrt{2m(V-E)}}{\hbar}.$$

Note that we have not discarded the exponentially increasing term in Ψ_{II}, as the barrier is finite and the wave-function does not have the opportunity to diverge to infinity.

Matching the wave-functions and their first derivatives at the boundaries ($x = 0$ and a) yields the following relationships:

$$A_1 + B_1 = A_2 + B_2,$$

$$ik_1A_1 - ik_1B_1 = k_2A_2 - k_2B_2,$$

$$A_2e^{k_2a} + B_2e^{-k_2a} = A_3e^{ik_1a},$$

$$k_2A_2e^{k_2a} - k_2B_2e^{-k_2a} = ik_1A_3e^{ik_1a}.$$

(2.29)

Solving these simultaneous equations, we obtain the transmission coefficient T:

$$T = \left|\frac{A_3}{A_1}\right|^2 = \frac{1}{1 + \left(\dfrac{k_1^2 + k_2^2}{2k_1k_2}\right)^2 \sinh^2(k_2a)}.$$

(2.30)

This is plotted as a function of incident electron energy in Fig. 2.18 for an example barrier 1 nm wide and 2 eV high. There is an exponential dependence of T on E.

What now about the case where $E > V$?

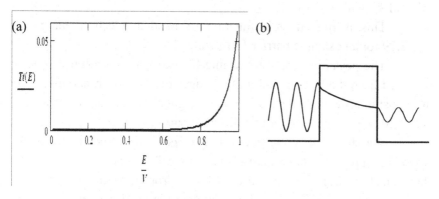

Fig. 2.18. (a) Transmission probability and (b) typical wave-function for a 1 nm wide, 2 eV high potential barrier, for the case where $E < V$ (tunnelling).

The solutions to Schrödinger's equation are of the form:

$$\Psi_I = A_1 e^{ik_1 x} + B_1 e^{-ik_1 x}, \quad \Psi_{II} = A_2 e^{ik_2 x} + B_2 e^{-ik_2 x}, \quad \Psi_{III} = A_3 e^{ik_3 x},$$

where $\quad k_1 = k_3 = \dfrac{\sqrt{2mE}}{\hbar}$ and $k_2 = \dfrac{\sqrt{2m(E-V)}}{\hbar}.$ (2.31)

Matching the wave-functions and their first derivatives at the boundaries ($x = 0$ and a) yields the following relationships:

$$A_1 + B_1 = A_2 + B_2$$

$$ik_1 A_1 - ik_1 B_1 = ik_2 A_2 - ik_2 B_2,$$

$$A_2 e^{ik_2 a} + B_2 e^{-ik_2 a} = A_3 e^{ik_1 a},$$

$$ik_2 A_2 e^{ik_2 a} - ik_2 B_2 e^{-ik_2 a} = ik_1 A_3 e^{ik_1 a}.$$

(2.32)

Solving these simultaneous equations, we obtain for the transmission coefficient T:

$$T = \left| \frac{A_3}{A_1} \right|^2 = \frac{1}{1 + \left(\dfrac{k_1^2 - k_2^2}{2k_1 k_2} \right)^2 \sin^2(k_2 a)}.$$ (2.33)

(This equation could have been arrived at simply by replacing the original k_2 with ik_2'.)

This is plotted as a function of incident electron energy in Fig. 2.19 for an example barrier 1 nm wide, 2 eV high.

The peaks in T are due to interference between electron waves scattered from the leading and trailing edges of the barrier, and therefore their positions depend on the width and height of the barrier, as shown in Fig. 2.20, which shows T for two different barrier widths.

We should now consider what happens to the potential profile when we apply a voltage across our "device". This is schematically shown in Fig. 2.21. There will be a linear drop in potential within the barrier region (as in the case of a capacitor) Fig. 2.21(a), which we will initially approximate to the stepped barrier structure in Fig. 2.21(b).

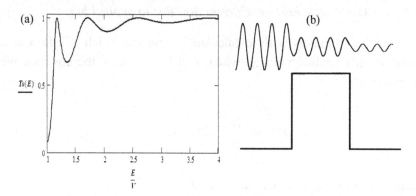

Fig. 2.19. (a) Transmission probability and (b) typical wave-function for a 1 nm wide, 2 eV high potential barrier, for the case where $E > V$ (scattering).

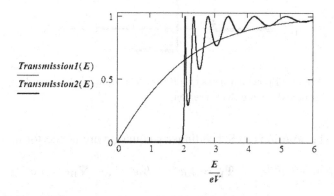

Fig. 2.20. Transmission probability for a 2 eV high potential barrier, for widths of 0.2 nm (thin line) and 2 nm (thick line, showing oscillatory behaviour).

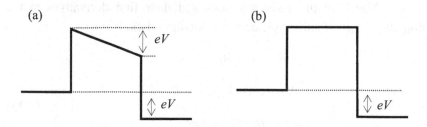

Fig. 2.21. Potential barrier under the application of a voltage, (a) actual potential profile, (b) approximation — stepped barrier.

2.4.3.2. *Asymmetric barrier: Current flow due to applied bias*

We ultimately wish to calculate the current which will flow as a result of this applied voltage, which will be given by the equation we derived earlier:

$$I^{1D} = \frac{2e}{h} \left(\int_{\mu_r}^{\mu_l} T(E)dE \right). \tag{2.11}$$

This is schematically shown in Fig. 2.22.

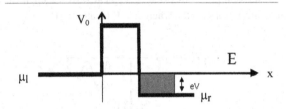

Fig. 2.22. A stepped potential barrier with a voltage, V applied across it. Current can flow with energies within the shaded region.

The solutions to Schrödinger's equation are of the form:

$$\Psi_{\mathrm{I}} = A_1 e^{ik_1 x} + B_1 e^{-ik_1 x}, \quad \Psi_{\mathrm{II}} = A_2 e^{k_2 x} + B_2 e^{-k_2 x}, \quad \Psi_{\mathrm{III}} = A_3 e^{ik_3 x},$$

$$\text{where } k_1 = \frac{\sqrt{2mE}}{\hbar}, \quad k_2 = \frac{\sqrt{2m(V_0 - E)}}{\hbar} \text{ and } k_3 = \frac{\sqrt{2m(E+V)}}{\hbar} \tag{2.34}$$

Matching the wave-functions and their first derivatives at the boundaries ($x = 0$ and a) yields the following relationships:

$$A_1 + B_1 = A_2 + B_2,$$

$$ik_1 A_1 - ik_1 B_1 = k_2 A_2 - k_2 B_2,$$

$$A_2 e^{k_2 a} + B_2 e^{-k_2 a} = A_3 e^{ik_3 a}, \tag{2.35}$$

$$k_2 A_2 e^{k_2 a} - k_2 B_2 e^{-k_2 a} = ik_3 A_3 e^{ik_3 a}.$$

Solving these simultaneous equations, we obtain the transmission coefficient T:

$$T = \frac{k_3}{k_1}\left|\frac{A_3}{A_1}\right|^2 = \left|\left(\frac{4k_1k_2}{2(k_1k_3 - k_2{}^2)\sinh(k_2a) + 2ik_2(k_1 + k_3)\cosh(k_2a)}\right)\right|^2 \frac{k_3}{k_1}.$$

$$(2.36)$$

This is plotted as a function of incident electron energy in Fig. 2.23 for an example barrier 1 nm wide, 2 eV high, and an applied bias of 0 and 1 V.

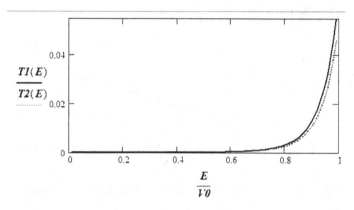

Fig. 2.23. Transmission probability for a stepped barrier for an applied bias of 0 V $(T_1(E))$ and 1 V $(T_2(E))$.

The applied voltage actually reduces the transmission probability, as it increases the amount of the potential discontinuity.

Now that we know the form of $T(E)$, we can use it to calculate the current through our stepped barrier structure. We are still assuming that this is a semiconductor device, i.e. we are tunnelling through a material rather than through a vacuum. Figure 2.24 shows the calculated current-voltage curve for a barrier which is 1 nm wide, 2 eV high, assuming there is one current channel open. Here, we have plotted the current up to a voltage of 0.2 V, which is 10% of the barrier height, as for larger voltages, the assumption of a stepped barrier structure starts to break down. This is of course a one-dimensional calculation, assuming one current mode. A realistic tunnel junction may have many modes, so the actual current may be significantly larger than the value we have calculated, scaling with the area of the device.

We can clearly see that in the low-voltage limit, the *I–V* characteristic of a tunnel barrier is linear.

From Eq. (2.35), we see that T depends very strongly on the barrier width, a. This is highlighted in Fig. 2.25, where we have calculated the current through a 2 eV high barrier, for an applied voltage of 0.2 V, as a function of barrier width from 0.24 nm (approximate interatomic spacing in metals, corresponding to a one monolayer thick quantum barrier) to 1 nm.

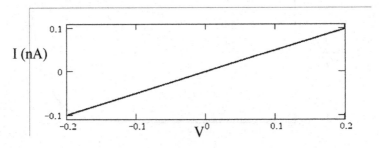

Fig. 2.24. Calculated current through a barrier of height 2 eV, width 1 nm.

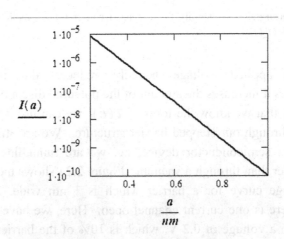

Fig. 2.25. Current through a barrier of height 2 eV, as a function of barrier width, for an applied voltage of 0.2 V.

The semi-log plot illustrates that the current depends inverse-exponentially on the barrier width, dropping by approximately an order of magnitude for each increase in barrier width of 0.1 nm. This characteristic is at the heart of the operating principle of the scanning tunnelling microscope, as we will see in Chapter 4.

We have also assumed that we are at zero Kelvin, so will have step-like Fermi functions. What about the effect of temperature? If we apply a voltage, V to a conductor, we shift the Fermi function to look like:

$$f(E) = \frac{1}{e^{\left(\frac{E - \mu + eV}{k_B T}\right)} + 1}.$$ (2.37)

Therefore, the expression for the current flow at non-zero temperature becomes:

$$J_{\text{total}} = \frac{2e}{\hbar}\left(\int_{\mu_r}^{\mu_l} \mathcal{D}(E)\left(f_l(E) - f_r(E_r)\right)T(E)dE\right)$$ (2.38)

$$J_{\text{total}} = \frac{2e}{\hbar}\left(\int_{\mu_r}^{\mu_l} \mathcal{D}(E)\left(\frac{1}{e^{\left(\frac{E - \mu}{k_B T}\right)} + 1} - \frac{1}{e^{\left(\frac{E - \mu + eV}{k_B T}\right)} + 1}\right)T(E)dE\right).$$ (2.39)

This is famously known as the "Tsu–Esaki formula" [16]. The effect of temperature can be seen in Fig. 2.26 where we have used this formula to compare the current–voltage characteristics at zero Kelvin and room temperature.

2.4.4. *Double potential barrier*

2.4.4.1. *Symmetric barriers: No applied voltage*

We shall now turn our attention to more complex barrier structures which have useful properties for real device applications, such as the symmetric double barrier shown in Fig. 2.27.

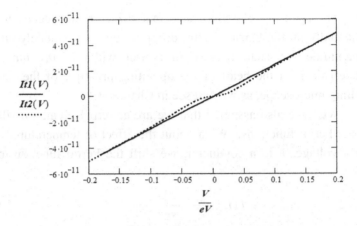

Fig. 2.26. Current through a barrier of height 2 eV, as a function of applied voltage, at 0 K (Solid line) and 300 K (dotted line).

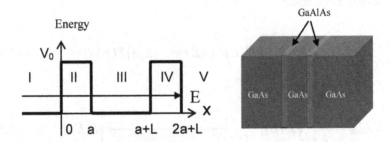

Fig. 2.27. (a) Double barrier structure; (b) construction of barrier.

The solutions to Schrödinger's equation in the five regions are of the form:

$$\Psi_{\mathrm{I}} = A_1 e^{ik_1 x} + B_1 e^{-ik_1 x},$$

$$\Psi_{\mathrm{II}} = A_2 e^{k_2 x} + B_2 e^{-k_2 x},$$

$$\Psi_{\mathrm{III}} = A_3 e^{ik_3 x} + B_3 e^{-ik_3 x}, \qquad (2.40)$$

$$\Psi_{\mathrm{IV}} = A_4 e^{k_4 x} + B_4 e^{-k_4 x},$$

$$\Psi_{\mathrm{V}} = A_5 e^{ik_5 x},$$

where $k_1 = k_3 = k_5 = \dfrac{\sqrt{2mE}}{\hbar}$ and $k_2 = k_4 = \dfrac{\sqrt{2m(V_0 - E)}}{\hbar}$.

We now have four boundaries at which to match our wave-functions and their derivatives, which gives the following relationships:

At $x = 0$ and a:

$$A_1 + B_1 = A_2 + B_2,$$

$$iA_1 k_1 - iB_1 k_1 = A_2 k_2 - B_2 k_2,$$

$$A_2 e^{k_2 a} + B_2 e^{-k_2 a} = A_3 e^{ik_1 a} + B_3 e^{-ik_1 a},$$

$$A_2 k_2 e^{k_2 a} - B_2 k_2 e^{-k_2 a} = ik_3 A_3 e^{ik_1 a} - ik_3 B_3 e^{-ik_1 a}.$$

(2.41)

At $x = a + L$ and $2a + L$:

$$A_3 e^{ik_1(a+L)} + B_3 e^{-ik_1(a+L)} = A_4 e^{k_2(a+L)} + B_4 e^{-k_4(a+L)},$$

$$ik_1 A_3 e^{ik_1(a+L)} - ik_1 B_3 e^{-ik_1(a+L)} = k_2 A_4 e^{k_2(a+L)} - k_2 B_4 e^{-k_4(a+L)},$$

$$A_4 e^{k_2(2a+L)} + B_4 e^{-k_2(2a+L)} = A_5 e^{ik_1(2a+L)},$$

$$k_2 A_4 e^{k_2(2a+L)} - k_2 B_4 e^{-k_2(2a+L)} = ik_1 A_5 e^{ik_1(2a+L)}.$$

(2.42)

We need to find $\left| A_5/A_1 \right|^2$, which we can determine using the relationships which follow from the above equations, where we have assumed $A_5 = 1$, so $T = \left| 1/A_1 \right|^2$.

If we write the above relationships in matrix form, we find the following:

$$\begin{pmatrix} 1 & 1 \\ ik_1 & -ik_1 \end{pmatrix}\begin{pmatrix} A_1 \\ B_1 \end{pmatrix} = \begin{pmatrix} 1 & 1 \\ k_2 & -k_2 \end{pmatrix}\begin{pmatrix} A_2 \\ B_2 \end{pmatrix},$$

$$\begin{pmatrix} e^{k_2 a} & e^{-k_2 a} \\ k_2 e^{k_2 a} & -k_2 e^{-k_2 a} \end{pmatrix}\begin{pmatrix} A_2 \\ B_2 \end{pmatrix} = \begin{pmatrix} e^{ik_1 a} & e^{-ik_1 a} \\ ik_3 e^{ik_1 a} & -ik_3 e^{-ik_1 a} \end{pmatrix}\begin{pmatrix} A_3 \\ B_3 \end{pmatrix},$$

$$\begin{pmatrix} e^{ik_1(a+L)} & e^{-ik_1(a+L)} \\ ik_1 e^{ik_1(a+L)} & -ik_1 e^{-ik_1(a+L)} \end{pmatrix}\begin{pmatrix} A_3 \\ B_3 \end{pmatrix} = \begin{pmatrix} e^{k_2(a+L)} & e^{-k_2(a+L)} \\ k_2 e^{k_2(a+L)} & -k_2 e^{-k_2(a+L)} \end{pmatrix}\begin{pmatrix} A_4 \\ B_4 \end{pmatrix},$$

$$\begin{pmatrix} e^{k_2(2a+L)} & e^{-k_2(2a+L)} \\ k_2 e^{k_2(2a+L)} & -k_2 e^{-k_2(2a+L)} \end{pmatrix}\begin{pmatrix} A_4 \\ B_4 \end{pmatrix} = \begin{pmatrix} e^{ik_1(2a+L)} & 0 \\ ik_1 e^{ik_1(2a+L)} & 0 \end{pmatrix}\begin{pmatrix} A_5 \\ B_5 \end{pmatrix}.$$

$$(2.43)$$

If we denote the matrices which prefix the wave-functions (i.e. A_n and B_n) as M_n, we obtain the following relationships:

$$M_1\begin{pmatrix} A_1 \\ B_1 \end{pmatrix} = M_2\begin{pmatrix} A_2 \\ B_2 \end{pmatrix} \Rightarrow \begin{pmatrix} A_1 \\ B_1 \end{pmatrix} = M_1^{-1} M_2\begin{pmatrix} A_2 \\ B_2 \end{pmatrix}. \qquad (2.44)$$

Carrying this through and considering all the barriers, we finally obtain

$$\begin{pmatrix} A_1 \\ B_1 \end{pmatrix} = M_1^{-1} M_2 M_3^{-1} M_4 M_5^{-1}\begin{pmatrix} A_5 \\ B_5 \end{pmatrix}. \qquad (2.45)$$

which can be re-written as:

$$\begin{aligned} A_1 &= M_{11} A_5 + M_{12} B_5, \\ B_1 &= M_{21} A_5 + M_{22} B_5. \end{aligned} \qquad (2.46)$$

The transmission coefficient is then given by $\left| M_{11}^{-1} \right|^2$, which can be determined readily using many commercially available mathematics packages.

This method of writing the equations in matrix form is known as the "*transfer matrix*" method and is extremely powerful.

When these equations are solved, we find a transmission probability as shown in Fig. 2.28.

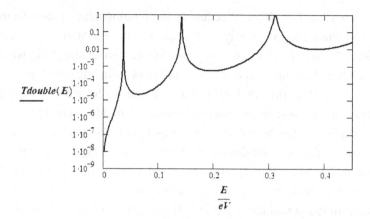

Fig. 2.28. Transmission probability for symmetric double barrier. Barriers are 3 nm wide, separated by 10 nm, and are 0.5 eV high.

For certain discrete values of incident electron energy, the transmission probability is almost exactly 1, which warrants further investigation, as the transmission probability for a single barrier of similar dimensions is significantly lower. In fact, the previous curve looks very similar to that expected for a resonant phenomenon. To understand this better, we should just briefly consider the quantum well which exists between the two barriers.

We can imagine that if we were to place an electron inside that well, it might only be able to exist there in one of a few "modes" or bound states.

As we saw earlier, a quantum well is directly analogous to an atom, wherein electrons reside in certain energy levels, or modes. It can be shown that any confining potential will always have at least one bound state associated with it. Figure 2.29 shows the bound states expected for a quantum well such as the one we have just considered, from which we can see there are three bound states at 0.038 eV, 0.149 eV and 0.323 eV, which are very close to the values of energy at which there are peaks in *T*, which are 0.036 eV, 0.141 eV and 0.31 eV, respectively. The slight disparity arises as the quantum well is not bounded by infinitely thick walls, but rather by ones which are only 3 nm thick. This has the effect of reducing the confinement of the electrons, so lowers their energy slightly with respect to the quantum well we have considered. In our calculations for the double-barrier structure, we have assumed the electron mass is that within GaAs, which is $0.067\,m_e$.

What does this analysis tell us? When we have a double-barrier structure, there are certain discrete values of incident electron energy, corresponding to the bound state energies within the well, for which the transmission through the device is almost 1. This type of behaviour is known as "resonant tunnelling", and what is occurring is that at those energies, the electron is exciting a metastable state within the well, and the wave-function maintains phase coherence throughout its transit.

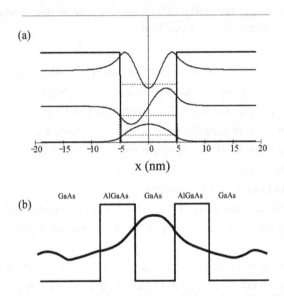

Fig. 2.29. (a) Bound states and energy levels within a GaAs/GaAlAs quantum well 10 nm wide and 0.5 eV deep; (b) wave-function for ground state of entire resonant tunnelling device.

2.4.4.2. *Tunnelling through multiple barriers with no phase coherence*

What about the case when there is scattering within the device and therefore there is no phase coherence, i.e. the distance between the barriers is at least several mfps? In that case, there can be no resonance, and the transmission through the device can be thought of as two sequential tunnelling events, first through barrier 1, with transmission probability T_1, and then through barrier 2 with transmission probability T_2. The total transmission probability through the device will not simply be $T_1 T_2$ however, as the electron wave will undergo multiple reflections between the barriers (with reflection probabilities R_1 and R_2), so the transmission is larger than first appearances suggest. After an electron wave with unit amplitude passes through the first barrier, the wave amplitude is T_1. This wave is partially reflected from the second barrier, and the amplitude of the backscattered wave is $T_1 R_2$. This wave reaches the first barrier again where an amount $T_1 R_1 R_2$ gets reflected. This continues ad infinitum, and therefore the total transmitted wave through the device is the sum of all these partially transmitted waves, as illustrated in Fig. 2.30.

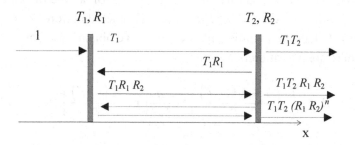

Fig. 2.30. Transmission through a double-barrier structure. Multiple reflection between barriers leads to enhanced transmission relative to $T_1 T_2$.

The sum of these waves is T which is given by:

$$T = T_1 T_2 \sum_{n=0}^{\infty} \left(R_1 R_2 \right)^n = \frac{T_1 T_2}{1 - R_1 R_2}. \tag{2.47}$$

As has been shown by Datta [7], remembering that $R_n = 1 - T_n$, the quantity

$$\frac{1-T}{T} = \frac{1-T_1}{T_1} + \frac{1-T_2}{T_2}, \tag{2.48}$$

which implies that if we had N scatterers in series, each having a transmission probability T_N, the net transmission probability through the N scatterers $T(N)$ is given by:

$$T(N) = \frac{1}{1 + \sum_N \dfrac{1-T_N}{T_N}}. \tag{2.49}$$

In the limiting case where each scatterer has an identical transmission probability, T we find that

$$T(N) = \frac{T}{T + N(1-T)}. \tag{2.50}$$

Earlier we found that the conductance of a one-dimensional ballistic conductor, $G^{1D} = (2e^2/h)T$. The resistance therefore is $R = 1/G$ $= (h/2e^2)/T$. Combining this with the above, we obtain for the resistance where there are N scatterers in series:

$$R = \frac{h}{2e^2}\left(\frac{T + N(1-T)}{T}\right) = \frac{h}{2e^2}\left(1 + \frac{N(1-T)}{T}\right). \tag{2.51}$$

The first term is independent of the conductor and represents the *contact resistance*, whereas the second term is due to the conductor itself. The resistance will scale inversely with the number of modes in the conductor, which we have assumed to just be one for now. If we take a closer look at this expression, we can write the resistance of the conductor as

$$R_{\text{cond}} = \frac{h}{2e^2}\frac{N(1-T)}{T}. \tag{2.52}$$

The number of scatterers within any material scales with its size, so for a conductor of length l, we can write $N = \gamma l$, where γ is the number of scatters per unit length of material. We can also say that the number of modes in a wire of cross-sectional area S is $S/(\lambda_F/2)^2$, where λ_F is the Fermi wavelength of the electrons. Combining this information, we obtain for the resistance of a conductor of cross-sectional area S and length l

$$R_{\text{cond}} = \frac{h}{2e^2}\left(\frac{\lambda_F}{2}\right)^2 \frac{(1-T)}{T}\frac{\gamma l}{S} = \frac{\rho l}{S}, \tag{2.53}$$

where ρ is the resistivity of the material, just as we saw in Chapter 1. Again, we have been able to use quantum mechanics to arrive at a macroscopic relationship.

The approach which we have just taken to calculate the total transmission probability is in fact more powerful than first appearances suggest. If we refine our calculation, and now consider what happens when the electron maintains phase coherence, there will be a phase shift of $\{2k_3L +$ the phase shift incurred by passing through the tunnel barriers$\}$ for each round-trip, which empirically gives a net transmission probability of approximately

$$T = T_1T_2 \sum_{n=0}^{\infty}\left(R_1R_2\right)^n e^{in\left(2Lk_3 + 0.95\frac{L+2a}{3}k_3\right)}. \tag{2.54}$$

This can be solved rather easily to give the approximate positions of the resonances for the case of an arbitrary number of barriers.

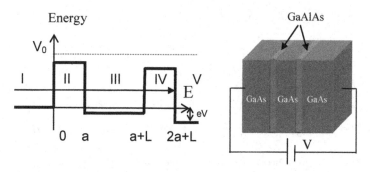

Fig. 2.31. (a) Double barrier structure with applied voltage, V; (b) construction of barrier.

2.4.4.3. *Asymmetric barriers: Applied voltage*

Returning to our double-barrier structure, what is the effect of applying a voltage across it? Keeping with our approximation that the voltage is dropped step-like across the barriers, the potential profile will look as shown in Fig. 2.31.

The solutions to Schrödinger's equation in the five regions are of the form:

$$\Psi_I = A_1 e^{ik_1 x} + B_1 e^{-ik_1 x}, \qquad \Psi_{II} = A_2 e^{k_2 x} + B_2 e^{-k_2 x}$$

$$\Psi_{III} = A_3 e^{ik_3 x} + B_3 e^{-ik_3 x} \qquad \Psi_{IV} = A_4 e^{k_4 x} + B_4 e^{-k_4 x} \qquad (2.55)$$

$$\Psi_V = A_5 e^{ik_5 x},$$

where $\quad k_1 = \dfrac{\sqrt{2mE}}{\hbar}, \quad k_2 = k_4 = \dfrac{\sqrt{2m(V_0 - E)}}{\hbar}, \quad k_3 = \dfrac{\sqrt{2m\left(E + \frac{V}{2}\right)}}{\hbar},$

and $\quad k_5 = \dfrac{\sqrt{2m(E + V)}}{\hbar}.$

And we have assumed that half of the applied voltage is dropped across the first barrier, and all across the second one.

Matching the wave-functions at the two boundaries as before leads to the following set of matrix equations:

$$\begin{pmatrix} 1 & 1 \\ ik_1 & -ik_1 \end{pmatrix} \begin{pmatrix} A_1 \\ B_1 \end{pmatrix} = \begin{pmatrix} 1 & 1 \\ k_2 & -k_2 \end{pmatrix} \begin{pmatrix} A_2 \\ B_2 \end{pmatrix},$$

$$\begin{pmatrix} e^{k_2 a} & e^{-k_2 a} \\ k_2 e^{k_2 a} & -k_2 e^{-k_2 a} \end{pmatrix} \begin{pmatrix} A_2 \\ B_2 \end{pmatrix} = \begin{pmatrix} e^{ik_3 a} & e^{-ik_3 a} \\ ik_3 e^{ik_3 a} & -ik_3 e^{-ik_3 a} \end{pmatrix} \begin{pmatrix} A_3 \\ B_3 \end{pmatrix},$$

$$\qquad (2.56)$$

$$\begin{pmatrix} e^{ik_3(a+L)} & e^{-ik_3(a+L)} \\ ik_3 e^{ik_3(a+L)} & -ik_3 e^{-ik_3(a+L)} \end{pmatrix} \begin{pmatrix} A_3 \\ B_3 \end{pmatrix} = \begin{pmatrix} e^{k_4(a+L)} & e^{-k_4(a+L)} \\ k_4 e^{k_4(a+L)} & -k_4 e^{-k_4(a+L)} \end{pmatrix} \begin{pmatrix} A_4 \\ B_4 \end{pmatrix},$$

$$\begin{pmatrix} e^{k_4(2a+L)} & e^{-k_4(2a+L)} \\ k_4 e^{k_4(2a+L)} & -k_4 e^{-k_4(2a+L)} \end{pmatrix} \begin{pmatrix} A_4 \\ B_4 \end{pmatrix} = \begin{pmatrix} e^{ik_5(2a+L)} & 0 \\ ik_5 e^{ik_5(2a+L)} & 0 \end{pmatrix} \begin{pmatrix} A_5 \\ B_5 \end{pmatrix},$$

which again gives us a similar relationship to before, i.e.

$$A_1 = M_{11}A_5 + M_{12}B_5,$$

$$B_1 = M_{21}A_5 + M_{22}B_5. \tag{2.57}$$

So the transmission coefficient is just $\left| M_{11}^{-1} \right|^2 k_5/k_1$.

Solving in the same way as for the symmetric double barriers, we obtain a transmission probability of the form shown in Fig. 2.32.

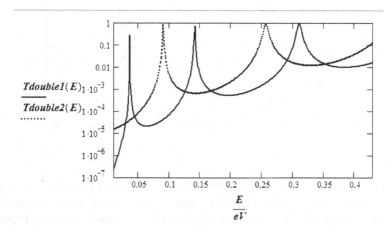

Fig. 2.32. Transmission probability for asymmetric double barriers with and without an applied voltage of 0.1 V (dotted and solid curves, respectively). Barriers are 3 nm wide, separated by 10 nm, in GaAs, and are 0.5 eV high.

In a similar manner to that seen for the single-barrier case, the transmission probability is modified through the application of a voltage across the device. We can use this in the same way as before to calculate the current–voltage characteristic for our device, as illustrated in Fig. 2.33.

This curve illustrates a phenomenon known as "Negative differential resistance", or NDR [17] — there are some situations whereby increasing the voltage across a device can cause the current to decrease, which are marked as "*A*" and "*B*" in Fig. 2.33. The peak current (i.e. at resonance *B*) is around 4 μA, whereas the off-resonance current (on the RHS of the peak) is around 1.5 μA. The ratio of the two, known as the peak–valley ratio is a measure of the quality of the device — the larger this ratio, the better the device is considered to be. The full-width at half-maximum of the resonance peak (ΔV) is related to the bandwidth of the energy level(s)/bands within the well (ΔE, in eV) in that $\Delta V = \Delta E$.

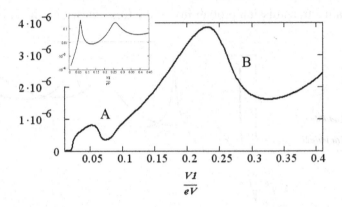

Fig. 2.33. Calculated current through double barrier structure of height 0.5 eV, barrier width 1 nm, barrier separation 10 nm, in GaAs. *A* and *B* denote regions of negative differential resistance (NDR). Inset shows $T(E,V)$. Peaks in current correspond to peaks in the transmission probability.

Also, the more bound states that exist within the well, the more peaks will be observed in the current–voltage characteristic, as illustrated in Fig. 2.34, which is a calculation for a 30 nm wide well.

One feature which is clear from these plots is that as the voltage is increased, the peak-width also increases. This is due to the fact that as we increase the voltage, the current is flowing through a wider range of energies, so all spectral features become smeared out. This is further compounded with increasing temperature, as shown in Fig. 2.35.

The effect of temperature in this particular case is quite noticeable in that the peaks are sharper at lower temperature. The reason for the greatly different magnitudes of current lie in the differences between the Fermi distribution at low and room temperature.

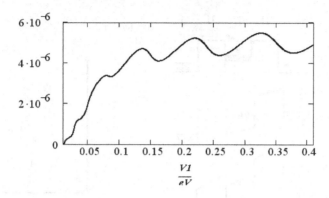

Fig. 2.34. Calculated current through double barrier structure of height 0.5 eV, barrier width 0.5 nm, barrier separation 30 nm, in GaAs.

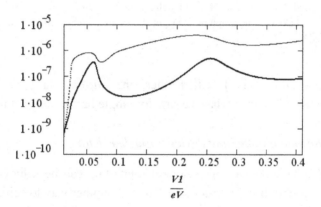

Fig. 2.35. Calculated current through double barrier structure of height 0.5 eV, barrier width 5 nm, barrier separation 10 nm for $T = 40$ K (solid curve) and 300 K (dotted curve).

We can describe the above current–voltage characteristics with the use of a simple band diagram, as illustrated in Fig. 2.36.

For no or a low applied voltage (regime A), very little current flows through the device. As soon as the voltage is large enough, as in regime B, the Fermi energy on the LHS overlaps with the energy of the bound states within the well, and there is a transmission resonance. For further increases of the voltage, as in regime C, there is no longer this overlap, so the current drops, and there is non-resonant tunnelling. As

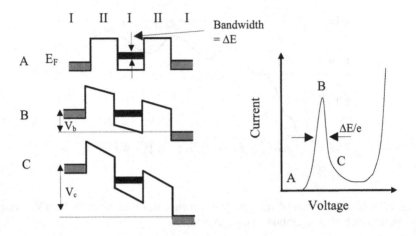

Fig. 2.36. Band diagram of resonant tunnelling diode, with 1 band of resonant levels.
I = GaAs, II = GaAlAs. The graph on the right shows the current–voltage relationship for
such a device.

the voltage is increased still further, the current starts to increase
exponentially, similar to what we saw for single barrier tunnelling.

2.4.4.4. *Resonant tunnelling devices: Further details*

We have shown that resonant tunnelling can be achieved when
we have a double-barrier structure. There is another way to achieve very
similar behaviour: if a conventional p–n junction diode is doped heavily
enough ($\sim 10^{25}$ dopants m^{-3}), it is possible to cause the Fermi levels in
the n and p-type materials to be in the conduction and valence bands,
respectively, as shown in Fig. 2.37. Also, the effect of very high doping
levels is to make the depletion region extremely thin, in the nm range, so
appreciable tunnelling can occur.

In Fig. 2.37(a), i.e. under zero applied bias: there is no net
current flow, as the electron current from the conduction band of the n-
type into the valence band of the p-type is balanced by the electron
current from the valence band of the p-type in to the conduction band of
the n-type. In Fig. 2.37(b), under reverse bias conditions, the bands on
the p-type side are raised relative to the n-type side, and electrons can

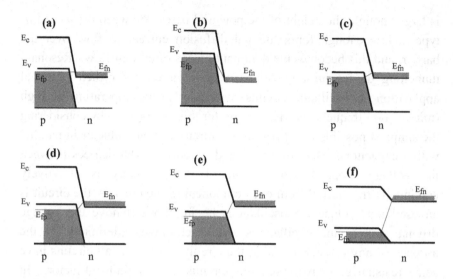

Fig. 2.37. Band diagram of tunnel diode.

flow from *p* to *n*, tunnelling across the depletion region. The width of this region will increase as the voltage is increased, so little current will actually flow. In Fig. 2.37(c), which is under a low forward bias, the electron-filled states in the *n*-type conduction band overlap with the holes in the *p*-type valence band and a significant current can tunnel across the depletion region, similar to Regime *B* in Fig. 2.37(b). In Fig. 2.37(d), as the forward bias is increased, the degree of overlap between the *n*-type conduction band electrons and the *p*-type valence band holes decreases, as more of them start to overlap with the band gap within the *p*-type. This has the effect of reducing the current across the depletion region as there are fewer states for the *n*-type electrons to tunnel into. In Fig. 2.37(e), similar to Regime *C* in Fig. 2.36(b), the current drops to its minimum value, as there is no longer any overlap between the conduction band electrons in the *n*-type and holes in the *p*-type: there are no available states for the electrons to tunnel into. The only current which can flow at this point is a small inelastic tunnel current and a small thermal diffusion current. In Fig. 2.37(f), when the applied forward bias is large enough, the height of the potential barrier between the *n*- and *p*-type is low enough for a thermal diffusion current to flow over the barrier, and this becomes the dominant means of current flow. Resonant tunnelling diodes initially gained a lot in interest for their potential application in oscillator circuits, particularly ones operating at high (microwave) frequencies. The reason for this can be seen by considering the simplest possible oscillator: an LC circuit (i.e. an inductor in parallel with a capacitor). Due to the phase difference of 180 degrees between the voltage dropped across each of these, energy is effectively continually transferred from one component to the other — the circuit is an oscillator. Once the oscillations begin, if we remove the voltage driving source, the oscillations would continue indefinitely in the absence of any resistance within the circuit. However, all circuits have some resistance, so real oscillator circuits have a finite Q-factor. In principle, if we could add a negative resistance into the circuit to counteract the stray resistance of the components, we could greatly increase the circuit's Q-factor. This is done by adding a resonant tunnelling diode into the LC circuit, and ensuring that it is operating in the middle of its NDR region. This is illustrated in Fig. 2.38. In recent years, the tunnel diode has been replaced by digital components which are more reliable and which have significantly better performance.

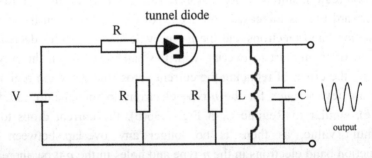

Fig. 2.38. Typical circuit utilising a tunnel diode. The voltage source V is used to set the diode operating in the NDR region (between A and B in Fig. 2.33(b)), and to start the oscillation. It also provides the energy to sustain the oscillation 3f the circuit. The oscillation frequency is $\left(\frac{1}{2\pi\sqrt{LC}} \right)$.

Where is this leading us? We have seen that the current through a device will depend on the applied voltage, as we know from our everyday experience. Quantum mechanics has unveiled that a factor relating current to voltage is the transmission probability, which for macroscopic systems varies so rapidly with energy on such a fine scale that it appears to be essentially independent of energy, whereas for nanometre sized structures, the variation as a function of energy is clearly manifest. For quantum structures smaller than the electronic mfp, we may observe complex current–voltage curves, with peaks (indicating resonant tunnelling), whereas for larger structures where the electronic phase coherence is smeared out, we will not observe any resonant characteristics unless we go to low temperatures.

2.4.5. *A more realistic calculation for a single potential barrier: The WKB approximation*

We saw earlier (Fig. 2.21) that the actual shape of the potential barrier becomes modified under the influence of an applied voltage, and becomes triangular, as shown below in Fig. 2.39.

The potential may be written as

$$V(x) = V_0 - eFx, \tag{2.58}$$

where F is the applied electric field.

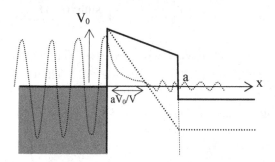

Fig. 2.39. Potential barrier under an applied electric field. Solid and dotted lines show potential profile when $V < V_0$ and $V > V_0$, respectively. Dotted curve shows a typical wave-function when $V > V_0$.

Qualitatively, what we would expect is that as the voltage is increased, there will come a point when the sloping part of the potential in the vacuum drops below V_0, and as the voltage is further increased, the effective width of the barrier will decrease. From this point onwards, the current will start increasing at a greater rate with increasing voltage, and significant currents can be drawn from conductors in this way. This phenomenon, which is called field emission is well-known and is commonly used in flat-panel displays, where each image pixel has a mini-tip located a short distance from it, and the pixel is turned on or off simply by applying a voltage between it and the mini-tip. The resulting field-emission current excites a phosphorescent coating on the pixel element.

To solve Schrödinger's equation for this form of potential is unfortunately non-trivial, and the solution is in the form of *Airy Functions*. Using approximation methods, it can however be solved in a straightforward manner. We are going to approach this using a very powerful approximation technique in quantum mechanics — the WKB (Wentzel, Kramers, Brilluoin) approximation [18]. Using this technique, we can treat complex tunnelling potential barriers in a piecewise manner. As a starting point, consider what we calculated earlier for the transmission probability of a symmetric tunnelling barrier:

$$T = \left| \frac{A_3}{A_1} \right|^2 = \frac{1}{1 + \left(\dfrac{k_1{}^2 + k_2{}^2}{2 k_1 k_2} \right)^2 \sinh^2(k_2 a)}.$$

If we consider the limiting case where $k_2 a \gg 1$ (i.e. the barrier is tall and thin), then $\sinh(k_2 a) = (e^{k_2 a} - e^{-k_2 a})/2 \sim e^{k_2 a}/2$. Therefore, $T \sim (4 k_1 k_2 / (k_1{}^2 + k_2{}^2))^2 e^{-2 k_2 a}$.

Taking the natural logarithm of T:

$$LnT \sim \ln(4 k_1 k_2 /(k_1{}^2 + k_2{}^2))^2 - 2 k_2 a \quad \sim -2 k_2 a \Rightarrow T \sim e^{-2 k_2 a}. \quad (2.59)$$

Therefore, if we take a barrier of arbitrary shape, split it up into a series of thin strips of width a, we can approximate the overall transmission probability as a sum over all those strips, i.e.:

$$T_n = e^{-2\sum_n k_n a_n}. \tag{2.60}$$

In the limit $a_n \to 0$, we can replace the summation with an integral,

$$T_n = e^{-2\int_A^B k_2(x)dx} = e^{-\frac{2}{\hbar}\int_A^B \sqrt{2m(V(x)-E)}dx}, \tag{2.61}$$

where A and B are the classical turning points, i.e. they define the region of space wherein the particle's energy is lower than that of the barrier, as shown in Fig. 2.40.

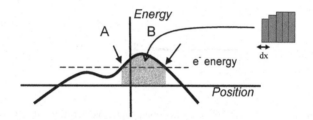

Fig. 2.40. General potential barrier with incident electron. Turning points denoted "A" and "B". Inset is zoom-in of barrier showing individual thin strips used in calculation.

If we insert $V(x)$ into this equation for T, we obtain

$$T = e^{-\frac{2}{\hbar}\int_A^B \sqrt{2m(V_0-eFx)}dx} = e^{-\frac{2\sqrt{2mV_0}}{\hbar}\int_A^B \sqrt{\left(1-eFx/V_0\right)}dx}. \tag{2.62}$$

We now need to determine the values of A and B. We have chosen $x = 0$ as the surface of the conductor, so that gives us $A = 0$. Simple geometry tells us that $B = a$ up to an electric field of (ϕ/ea), and for larger fields, $B = V_0/eF$. This is plotted in Fig. 2.41, and the current in Fig. 2.42.

The current increases exponentially with increasing voltage, at a rate determined by the initial barrier height, as is shown in Fig. 2.43.

We can solve the above equation for T analytically, which gives us:

$$T = \exp\left(-\tfrac{4}{3eF\hbar}\sqrt{2m}V_0^{\frac{3}{2}}\right). \tag{2.63}$$

This is a simplified version of the *"Fowler–Nordheim equation"* [19], which is commonly used to describe field-emission. Experimentally, a

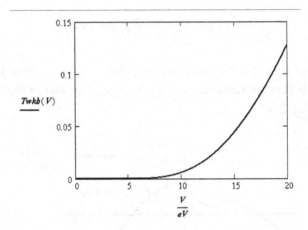

Fig. 2.41. Transmission probability for a sloped single barrier of width 1 nm and height 4 eV, as calculated using the WKB approximation.

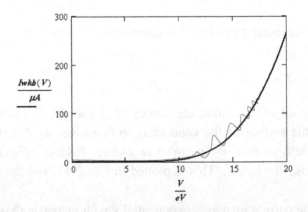

Fig. 2.42. Current through a sloped single barrier of width 1 nm and height 4 eV, as calculated using the WKB approximation. The dotted curve indicates (schematically) what is observed experimentally, and what a full calculation shows — Gundlach oscillations.

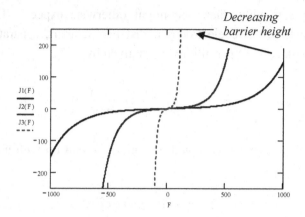

Fig. 2.43. Current through a sloped single barrier of width 1 nm and height 4 eV, as calculated using the WKB approximation.

high voltage is applied between a material and a collector, and the resulting current–voltage curves are fitted to this equation, from which the barrier height (in most cases, this is the work-function of the conductor) and the electric field enhancement factor can be calculated. The latter is a function of geometry, and is typically the same as the aspect ratio of the field-emitter — a long, sharp rod will emit more efficiently than a short, fat one. As the WKB approximation essentially ignores scattering-type transmission, it predicts a smooth, exponential rise in current with increasing voltage. Experimentally, this is not exactly the case, as the current is seen to also have an oscillatory component, as seen in Fig. 2.42 (these are known as Gundlach oscillations [20]). This can be analysed theoretically using the full Airy Functions for the wave-functions.

    In the case of vacuum tunnelling (or tunnelling through an insulator), we should be aware of the fact that whilst the electron is in the tunnel gap, it will induce image charges in the two electrodes. This serves to modify the barrier potential. The net effect of this is to reduce the average barrier height and hence increase the transmission probability. Because of the image potential (and the associated image force) [21], the current through a tunnel junction increases faster with decreasing gap width than one might otherwise expect. The image potential between two electrodes a distance s, apart, separated by an insulator with relative permitivity ε_r is given by

$$V_{\text{image}}(x) = -\frac{q^2}{4x} - \frac{q^2}{8\pi\varepsilon_0\varepsilon_r}\sum_{m=1}^{\infty}\left\{m\frac{s}{(ms)^2-x^2}-\frac{1}{ms}\right\}. \quad (2.64)$$

Combining this with the original potential, we end up with a combined barrier potential of

$$V(x) = \varphi + V_{\text{image}}(x) - eFx, \quad (2.65)$$

which looks as shown in Fig. 2.44.

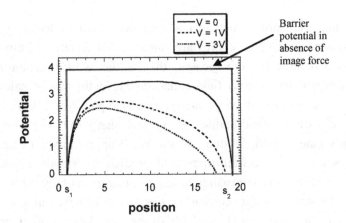

Fig. 2.44. Potential between two metal electrodes, including the image potential, for electrodes 20 Angstroms apart. Both electrodes have a work-function of 4 eV.

This problem has already been tackled in detail by Simmons [22], so we just reproduce his results here, for the case where the image potential is replaced by a parabolic potential, which is simpler to handle analytically.

The current density through a tunnel junction for the case of low applied voltage (i.e. $V \sim 0$) is given by:

$$J = \frac{3.16 \times 10^{10}}{\Delta s} \sqrt{\varphi_L} V e^{-1.025 \Delta s \sqrt{\varphi_L}},$$

where $\qquad \varphi_L = \left(\varphi_0 - \left\{ \frac{5.75}{\sqrt{\varepsilon_r} \left(s_2 - s_1 \right)} \right\} \ln \left[\frac{s_2 \left(s - s_1 \right)}{s_1 \left(s - s_2 \right)} \right] \right),$ \qquad (2.66)

$$s_1 = \frac{6}{\sqrt{\varepsilon_r} \, \varphi_0} \quad \text{and} \quad s_2 = s - \frac{6}{\sqrt{\varepsilon_r} \, \varphi_0}.$$

As before, s_1 and s_2 are the classical turning points, i.e. $\Delta s = (s_1 - s_2)$ for low voltages. The units of these equations are: J is in A. m^{-2}, $\phi_{0,L}$ are in eV, and s, s_1 and s_2 are in Angstroms.

The current density through a tunnel junction for the general case of *all* applied voltages is given by:

$$J = \frac{6.2 \times 10^{10}}{\Delta s^2} \left\{ \varphi_1 e^{-1.025 \Delta s \sqrt{\varphi_1}} - \left(\varphi_1 + V \right) e^{-1.025 \Delta s \sqrt{\varphi_1 + V}} \right\},$$

(2.67)

where $\varphi_1 = \left(\varphi_0 - V \frac{s_1 + s_2}{2s} - \left\{ \frac{5.75}{\sqrt{\varepsilon_r} \left(s_2 - s_1 \right)} \right\} \ln \left[\frac{s_2 \left(s - s_1 \right)}{s_1 \left(s - s_2 \right)} \right] \right).$

When $V < \varphi_0$, $s_1 = \frac{6}{\sqrt{\varepsilon_r} \, \varphi_0}$,

and $\quad s_2 = s \left(1 - \frac{46}{20 + 3 \varphi_0 s \sqrt{\varepsilon_r} - 2Vs \sqrt{\varepsilon_r}} \right) + \frac{6}{\sqrt{\varepsilon_r} \, \varphi_0}$,

and when $V > \varphi_0$, $s_1 = \frac{6}{\sqrt{\varepsilon_r} \, \varphi_0}$ and $s_2 = \frac{s \varphi_0 \sqrt{\varepsilon_r} - 28}{\sqrt{\varepsilon_r} V}$.

These equations, i.e. (Eqs. (2.67) and (2.68)) are plotted in Fig. 2.45 for a barrier of nominal height 4 eV, and gap widths between 0.8 nm and 2 nm, which are typical of the tip–sample distance in a scanning tunnelling microscope (see Chapter 4).

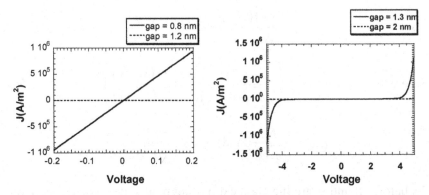

Fig. 2.45. Tunnel current density between two metal electrodes, including the image potential. Both electrodes have a work-function of 4 eV; (a) low-voltage limit, plotted using Eq. (2.67), for electrodes 8 and 12 Angstroms apart, (b) general voltage, Eq. (2.68), for electrodes 13 and 20 Angstroms apart. These are using the parabolic potential WKB approximation, so do not show Gundlach oscillations, which an exact calculation shows (see Fig. 2.42).

2.5. Techniques for the Fabrication of Quantum Nanostructures

There are now several texts which describe in detail the techniques used to experimentally fabricate nanostructures [23–25], so we will not devote much time to this here, apart from a brief overview. There are two ways in which materials with nm dimensions can be made artificially (i.e. as opposed to synthesising molecules): either by vertical or lateral patterning. Vertical patterning is by far the simplest, as it is possible to deposit materials by evaporation with sub-monolayer precision. This is the method used to fabricate all the device structures so far encountered in this text, which all only have confinement in one direction. The technique used to deposit GaAs, AlGaAs etc. is called Molecular-beam Epitaxy, or MBE for short [26]. This essentially consists of an ultra-high vacuum (UHV) chamber with evaporation sources, thickness monitors, sample heaters and a rotating sample holder to increase film uniformity. As the growth process is epitaxial (i.e. there is negligible lattice mis-match between the substrate and the deposited layers), the deposited layers are single-crystal with very few defects and the interfaces are atomically sharp. Both of these lead to excellent electrical device properties (the electrons have a very high mobility and long mfp due to the high level of crystalline perfection). The techniques which may be employed for lateral patterning are much more complex, but briefly, these techniques broadly fall into two classes: *top-down* and *bottom-up*. In top-down nanofabrication, one generally starts with a piece of material and progressively removes bits, to eventually be left with the desired structure, a little like sculpting. The techniques used to remove bits of unwanted material are collectively known as lithography. Conventional microprocessors are made using multi-step top-down optical lithography, where light is shone through a mask onto photosensitive material on the substrate. This is a highly parallel process: a large number of structures can be patterned simultaneously, at a resolution which is determined by the wavelength of the illumination source, but is typically below 100 nm. At a research level, devices and nanostructures are often made by top-down electron-beam (usually just called "e-beam") lithography. This is a serial process, so is significantly slower than optical lithography, but it has a better resolution of a few nm. There are a variety of different resists used in optical lithography with different properties depending on the dimensions of the structure being fabricated. In e-beam lithography, the most commonly used resist is PMMA (poly methyl methacrylate). As with photography, lithography can usually operate in either a positive or negative sense, i.e. the end structure can either have the shape of the mask or its inverse, as shown in Fig. 2.46.

Once the lithography has been done and the resist has been developed, there are a number of ways to deposit the material of interest, but the most common method is thermal evaporation, followed by lift-off.

The other way in which nanostructures can be made, and what has perhaps gained the most popular attention is the so-called "bottom-up" approach. In this way, nanostructures are basically *grown* up from the substrate, molecule-by-molecule or even atom-by-atom. One of the most aesthetically beautiful examples of bottom-up fabrication is the quantum coral shown in Fig. 2.4, where 56 iron atoms were moved using an STM tip to form a perfect circle on a copper surface. There are now

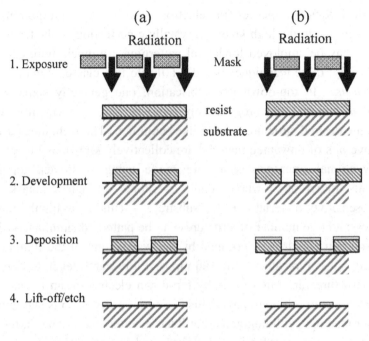

Fig. 2.46. Top-down lithography: (a) positive process, (b) negative process.

many examples of this sort of structure to be found in the literature, made using atoms or molecules, which are an excellent playground for surface physics experiments. Of course there is a need to be able to fabricate large numbers of structures and devices rather than just one-off ones, hence the interest in the process known as "self-assembly" [27]. Self-assembly is simply the fact that when some molecules are placed on certain surfaces (mostly molecules with thiol end-groups, and gold or copper surfaces) they tend to form close-packed and well-ordered monolayers. In combination with conventional lithographic techniques, it is possible to induce molecules to line up on predetermined areas of a sample. Further complexity can be achieved through the simultaneous use of several different molecules. In this way, many different structures may be realised. Perhaps the most important self-assembly is the formation of DNA. One of the simplest forms of self-assembly is the formation of lipid micelles, as shown below in Fig. 2.47.

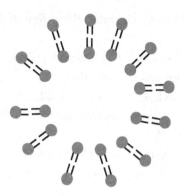

Fig. 2.47. Self-assembly — lipid micelle. Hydrophobic tails (black lines) are shielded from the outside world by the hydrophilic heads (grey spheres), and by the fact that the membrane can fold up into a closed surface. These structures form spontaneously.

References for Chapter 2

1. R. Feynman, *J. Microelectromechanical Systems* 1, 60 (1959).
2. US Patent #02569347
3. T. Sato, H. Ahmed, D. Brown and B. F. G. Johnson, *J. Appl. Phys.* 82, 696 (1997).
4. G. Binnig and H. Rohrer, *Helv. Phys. Acta* 55, 726 (1982).
5. M. F. Crommie, C. P. Lutz and D. M. Eigler, *Science* 262, 218 (1993).
6. N. W. Ashcroft and N. D. Mermin, *Solid State Physics* (Saunders College Publishing, Philadelphia, 1976).
7. S. Datta, *Electronic Transport in Mesoscopic Systems* (Cambridge University Press, Cambridge, 1995).
8. R. G. Mazur and D. H. Dickey, *J. Electrochemical Soc.* 113, 255 (1966).
9. Y. V. Sharvin, *Sov. Phys. JETP* 21, 655 (1965).
10. U. Landman, W. D. Luedtke, B. E. Salisbury and R. Whetten, *Phys. Rev. Lett.* 77, 1362 (1996).
11. B. E. Kane, G. R. Facer, A. S. Dzurak, N. E. Lumpkin, R. G. Clark, L. N. Pfeiffer and K. W. West, *Appl. Phys. Lett.* 75, 3506 (1998).
12. G. Rubio, N. Agrait and S. Vieira, *Phys. Rev. Lett.* 76, 2302 (1996).

13. D. J. BenDaniel and C. B. Duke, *Phys. Rev.* **152**, 683 (1966).
14. GaAs, etc.
15. A. I. M. Rae, *Quantum Mechanics*, 2nd edn. (Adam Hilger Publishing, Bristol and Boston, 1986).
16. R. Tsu and L. Esaki, *Appl. Phys. Lett.* **22**, 562 (1973).
17. M. J. Kelly, *Low Dimensional Semiconductors* (Oxford University Press, New York, 1995).
18. L. Schiff, *Quantum Mechanics* (McGraw Hill, New York, 1968).
19. R. H. Fowler and L. Nordheim, *Proc. Roy. Soc. Lond.* **A119**, 173 (1928).
20. K. H. Gundlach, *Solid-State Electron.* **9**, 946 (1966).
21. J. D. Jackson, *Classical Electrodynamics*, 3rd edn. (Wiley, 1998).
22. J. G. Simmons, *J. Appl. Phys.* **34**, 1793 (1963).
23. G. Cao, *Nanostructures and Nanomaterials: Synthesis, Properties and Applications* (Imperial College Press, 2004).
24. Scientific American, *Understanding Nanotechnology* (Warner Books, 2002).
25. C. P. Poole, Jr. and F. J. Owens, *Introduction to Nanotechnology* (Wiley, 2003).
26. M. A. Herman and H. Sitter, *Molecular Beam Epitaxy: Fundamentals and Current Status*, Springer Series in Materials Science (Springer, 1997).
27. G. M. Whitesides, J. P. Mathias and C. T. Seto, *Science* **254**, 1312 (1991).

Problems for Chapter 2

1. A ball of mass 1 gram is moving at 150 m/s. A 1000 kg automobile is moving at 90 km/hour. An oxygen molecule in air is moving at 10 m/s. What is the kinetic energy (in joules and in electron-volts), de Broglie wavelength in meter, and Planck frequency in Hz of these moving objects? Which particle may require wave-quantum mechanics to explain its motion and why?

2. In a television, we have a beam of electrons at 500 V. They are passed through a hollow metal cylinder which is at a potential of 200 V. If we can assume that this cylinder is infinitely long, what proportion of the incident beam will be reflected at its entrance? What will happen if the cylinder is in fact only 2 nm long?

3. If we swap the voltages to the beam and the 2 nm long cylinder, estimate the proportion of the incident beam which passes through the cylinder now.

4. Show that if an electron is placed in any symmetric potential well, then the corresponding wave-functions must be either symmetric or anti-symmetric.

5. Consider a quantum well in GaAs which is 5 nm wide, and which has electron and hole effective masses of 0.3 m_e, and 0.06 m_e, respectively, where m_e is the free-electron mass. Given that the band-gap of the semiconductor in the quantum well is 1.2 eV, sketch the approximate form of the optical density of the well in the range 0–1.6 eV, assuming that light will be strongly absorbed whenever the photon energy matches the energy of a transition between any levels in the conduction band and the valence band.

Chapter 3

Mesoscopic Transport:
Between the Nanoscale and the Macroscale

3.1. Introduction

The scale of interconnects used in the semiconductor industry is continually shrinking towards dimensions comparable with the electronic mean free path (mfp). The electrical transport properties of macroscopic wires are well-established, the resistance (Ω) following the simple relationship $\Omega = \rho l / A$, where ρ is the resistivity, and l and A are the sample length and cross-sectional area, respectively. We have already seen that wires with dimensions comparable with the Fermi wavelength (λ_f), through confinement of the electronic wave-functions by the surfaces exhibit discrete resistance values, given by $\Omega \propto 1/\text{Int}[A/\lambda_f^2]$, showing a step-wise variation with size. Transport at this scale is well-described using the highly successful Landauer–Büttiker formalism [1]. We saw in the last chapter that Landauer derived a formula relating the current through a structure to its quantum mechanical transmission probability, T. Büttiker extended this to the case of multiple inputs and outputs to the system (electrodes).

The intermediate region where a wire has dimensions of the order of the mfp is however a less well studied area and is where we briefly turn our attention to in this chapter. Extensive research has been carried out on extended thin films where there is confinement in only one dimension, and we will draw on the same tools used to study those systems in order to understand the effect of an extra degree of confinement, in the form of a wire. In this chapter we will take a brief look at the mechanisms of surface and grain-boundary scattering in the cases of thin films and wires.

3.2. Boltzmann Transport Equation

We have assumed up to now that conductivity is determined by the relaxation time of electrons. However, this approach starts to break down when there is surface and/or grain-boundary scattering as then the *effective* mfp depends on the conductor's geometry. The concept of a single relaxation time is then invalid and we need to independently consider the effective relaxation times due to each of the different sources of scattering. The way in which we consider the effect of geometry is to use the Boltzmann transport equation (which is an equation used to describe the dynamics molecules in a gas — in this case the electron gas inside a conductor) [2,3]. This equation is used whenever we must consider that the electron current density in a conductor varies with position, and has the form:

$$v \cdot \frac{\partial n}{\partial r} - \frac{eE_x}{m} \frac{\partial N_0}{\partial v_x} - \frac{n}{\tau}, \tag{3.1}$$

where v is the electron velocity, an electric field E_x is applied in the x-direction, N_0 is the electron distribution function, and n is the change in the number of electrons travelling in the x-direction due to the applied electric field. A major strand of mesoscopic transport physics is concerned with finding solutions to the Boltzmann transport equation for different situations, in order to predict and understand the resistance of device structures.

3.3. Resistivity of Thin Films and Wires: Surface Scattering

3.3.1. *General principles*

It is well known that the electrical resistivity of thin metallic films increases once the film thickness decreases below the bulk electronic mfp. This is often called the *Fuchs size effect* after initial work by Fuchs and Sondheimer (which we will denote by F–S theory) [4,5] which attributed this effect to diffuse scattering at the film boundaries essentially imposing a restriction on the mfp, as shown in Fig. 3.1.

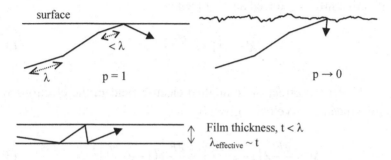

Fig. 3.1. Illustration of surface scattering and effect of film surface on mfp.

As resistivity is inversely proportional to the mfp, the resistivity consequently increases as a film gets thinner, below the bulk mfp. Their analysis consisted of solving the Boltzmann transport equation subject to the condition that at the film surfaces, a proportion of the electron distribution function is independent of direction (diffuse scattering). They found reasonable agreement with experimental results for thin Al and Sn films. Their work was extended to the case of wires of square [6], circular [7] and finally arbitrary [8] cross-section, with confinement now in two directions.

A simpler and more flexible approach due to Chambers [8], and based on using kinetic-theory arguments rather than solving the Boltzmann equation explicitly is the approach we take here. In the context of this analysis, the only unknown parameter is p, the proportion of electrons which are specularly reflected at the film surfaces. Until relatively recently, the standard procedure was to fit experimental resistivity/film-thickness data using p as the variable parameter.

Initially, we consider the case of fully diffuse surface scattering ($p = 0$), i.e. the electrons have a randomly oriented momentum after collision with the surface. Consider an electron starting at a point, I within the conductor, moving in the direction IP, where P is a point on the surface. For a mfp denoted by λ, the probability of the electron travelling a distance x (where $x < IP$) without scattering is $e^{-x/\lambda}$. In general, the mean distance travelled by an electron before scattering in a conductor is just λ, but in this case where there are surfaces to consider, the effective mfp is reduced and is given by

$$\lambda' = \lambda\left(1 - e^{-IP/\lambda}\right). \tag{3.2}$$

Under the effect of an applied electric field E, the electrons will acquire a mean drift velocity given by

$$\Delta v = \frac{eE}{mv}\lambda\left(1 - e^{-IP/\lambda}\right) = \frac{eE}{m}\tau\left(1 - e^{-IP/\lambda}\right). \tag{3.3}$$

The current will depend on the change in the number of electrons travelling in the direction IP driven by the electric field, which is in the x-direction. This is given by

$$n(IP) = \frac{\partial N_0}{\partial v_x}\Delta v_x = \frac{\partial N_0}{\partial v_x}\frac{eE_x}{m}\tau\left(1 - e^{-IP/\lambda}\right). \tag{3.4}$$

The current through any point is the sum of electrons travelling at all possible speeds in all possible directions through this point, or

$$i = \int_0^\infty v^2 dv \int_0^{2\pi} d\phi \int_0^\pi ev_x n(PI)\sin\theta d\theta, \tag{3.5}$$

$$i = \frac{e^2 E_x \tau}{m}\int_0^\infty v^3 \frac{\partial N_0}{\partial v} dv \int_0^{2\pi} d\phi \int_0^\pi \left(1 - e^{-IP/\lambda}\right)\sin\theta\cos^2\theta d\theta, \tag{3.6}$$

where the $\cos^2\theta$ term has come from replacing v_x with $v\cos\theta$ and $\partial N_0/\partial v_x$ with $\partial N_0/\partial v \partial v/\partial v_x$.

The total current passing through the film is found by integrating over the cross-sectional area of the film, to take account of different starting points, I; the conductivity is current density/applied field, which gives for the conductivity

$$\sigma = \frac{e^2 \tau}{ms} \int_0^\infty v^3 \frac{\partial N_0}{\partial v} dv \int_s ds \int_0^{2\pi} d\phi \int_0^\pi \left(1 - e^{-IP/\lambda}\right) \sin\theta \cos^2\theta d\theta. \qquad (3.7)$$

Now, the conductivity of a bulk sample is given by

$$\sigma_0 = \frac{4\pi}{3} \frac{e^2 \tau}{m} \int_0^\infty v^3 \frac{\partial N_0}{\partial v} dv, \qquad (3.8)$$

which gives for the ratio of conductivities of a thin film and a bulk sample:

$$\frac{\sigma}{\sigma_0} = \frac{\rho_0}{\rho} = \frac{3}{4\pi s} \int_s ds \int_0^{2\pi} d\phi \int_0^\pi \left(1 - e^{-IP/\lambda}\right) \sin\theta \cos^2\theta d\theta, \qquad (3.9)$$

$$\frac{\sigma}{\sigma_0} = \frac{\rho_0}{\rho} = 1 - \frac{3}{4\pi s} \int_s ds \int_0^{2\pi} d\phi \int_0^\pi e^{-IP/\lambda} \sin\theta \cos^2\theta d\theta. \qquad (3.10)$$

All that remains is to evaluate this by calculating IP in terms of the geometry and the position of the starting point, I.

3.3.2. 1D confinement: Thin film

For the case of a thin film, the above integral reduces to

$$\frac{\rho_0}{\rho} = 1 - \frac{3}{2t} \int_0^t dx \int_0^\pi e^{-t/\lambda \sin\theta} \sin\theta \cos^2\theta d\theta. \qquad (3.11)$$

For the more general case where we have partially specular reflection of electrons from the surface, the resistivity ratio is modified as follows [8]:

$$\frac{\rho_0}{\rho} = (1-p)^2 \sum_{n=1}^{\infty} \left\{ np^{n-1} \left(\frac{\sigma}{\sigma_0} \right)_{p=0,\lambda/n} \right\}. \tag{3.12}$$

It is relatively straightforward to evaluate these formulae using numerical techniques, and doing so, we have arrived at the plots in Fig. 3.2. This shows the calculated dependence of resistivity on film thickness for a gold thin film for entirely diffuse scattering and also for 50% diffuse scattering ($p = 0.5$).

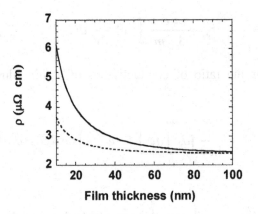

Fig. 3.2. Resistivity (surface-scattering component) versus film thickness for gold films where $\lambda = 40$ nm; solid curve, $p = 0$, dotted curve, $p = 0.5$.

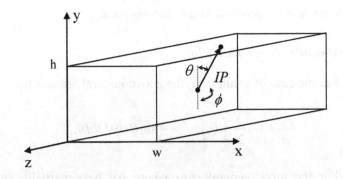

Fig. 3.3. Geometry for calculating surface scattering component of resistivity in thin wires.

3.3.3. *2D confinement: Rectangular wire*

For the case of a wire of rectangular cross-section where the wire width and thickness are w and t respectively, the length IP of the point I with co-ordinates (x,y) where $(0,0)$ is at the centre of the wire, is given by $x' = x/(\sin\theta\cos\phi)$, $y' = (y-h)/(\sin\theta\cos\phi)$, as illustrated in Fig. 3.3. The resistivity ratio is then given by

$$\frac{\rho_0}{\rho} = 1 - \frac{6}{4\pi hw}\left\{\int_0^w dx \int_0^t dy \int_{\arctan(y/x)}^{\arctan((h-y)/x)} d\phi \int_0^\pi e^{-x/\lambda\sin\theta\cos\phi}\sin\theta\cos^2\theta d\theta \right.$$

$$\left. + \int_0^w dx \int_0^t dy \int_{\arctan(x/(h-y))}^{\arctan((w-x)/(h-y))} d\phi \int_0^\pi e^{-(h-y)/\lambda\sin\theta\cos\phi}\sin\theta\cos^2\theta d\theta \right\}. \quad (3.13)$$

This expression may also be evaluated numerically, although to do so is more computationally intensive than in the case of a thin film, as now there are four integration variables. In Fig. 3.4, we have plotted the resistivity versus width for a 20 nm thick wire, for both fully diffuse and partially diffuse surface scattering.

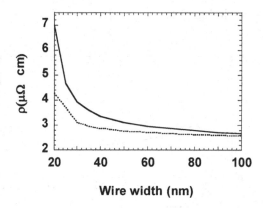

Fig. 3.4. Resistivity (surface scattering component) versus wire width for 20 nm thick gold nanowires where $\lambda = 40$ nm; solid curve, $p = 0$, dotted curve, $p = 0.5$.

3.3.4. *2D confinement: Cylindrical wires*

The case of a wire of cylindrical geometry is intermediate in that there are only three integration variables involved: the azimuthal angle, ϕ, the distance from a point (x, ϕ) to the surface, $(a-x)$, and the radius of the wire, a. This is illustrated in Fig. 3.5.

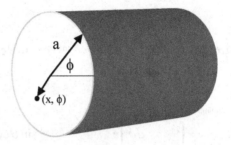

Fig. 3.5. Cylindrical wire of radius a, for resistivity calculations.

In this case, Eq. (3.10) reduces to (after Dingle [7]):

$$\frac{\rho_0}{\rho} = 1 - \frac{3}{\pi a^2} \int_{-a}^{a} dx \int_{-\sqrt{a^2-x^2}}^{\sqrt{a^2-x^2}} dy \int_{0}^{\pi/2} e^{\left(\frac{-y-\sqrt{a^2-x^2}}{\lambda \sin\phi}\right)} \sin\phi \cos^2\phi \, d\phi. \qquad (3.14)$$

This is plotted below in Fig. 3.6.

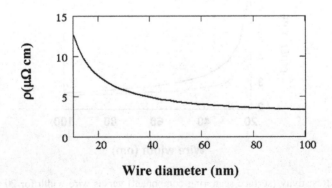

Fig. 3.6. Resistivity (surface scattering component) versus wire diameter for a gold wire.

3.4. Resistivity of Thin Films and Wires: Grain-Boundary Scattering

Initial measurements in the 1940s and 1950s on thin films and wires showed very good agreement with the above models. However, in those early days it was not possible to make very thin films or narrow wires, so most of those experiments were with micron-scaled wires, and low temperatures were needed to increase the mean free path to be comparable to the film or wire dimensions.

Towards the end of the 1960s however, significant departures from this model were found [9] when technological advances made it possible to fabricate smaller wires and thinner films. The situation was partially resolved by theoretical work done by Mayadas and Shatzkes (which we will denote as M–S theory) [10,11], who attributed the enhanced resistivity of thin films to grain-boundary scattering superimposed on the Fuchs size effect. The key to their work lay in the observation that up to film thicknesses of the order 1 μm, the mean film grain diameter is approximately equal to the film thickness. Consequently, thinner films tend to have smaller grains and hence more grain boundaries, leading to an increased resistivity. In their analysis, the resistivity due to grain boundary scattering is found to greatly exceed that due to surface scattering. The matter of grain-boundary versus surface scattering remained somewhat unresolved however until recently, as both theories provide a reasonable fit to experimental data for a wide variety of cases. The main parameters of the M–S theory are D_{50} the mean grain size, and the electron reflection coefficient R, which is the mean probability for an electron to be reflected by a grain boundary. Gold is known to exhibit a high degree of specular reflection from its bare surface, although interfaces with other metals can decrease this. From data fits to the M–S theory, one can infer $R \sim 0.15$ for Al, and values measured by transport measurements and STM potentiometry for R for single grain boundaries in Au yield R from 0.4 to as high as 0.9 [12–14]. One can estimate R quite readily by assuming that the grain boundary is equivalent to a missing row of atoms, and consequently the electronic barrier height is reduced significantly below the vacuum level due to the image potential, as illustrated in Fig. 3.7. Using the WKB

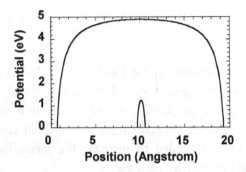

Fig. 3.7. Potential of electron in the vicinity of a grain boundary in gold, for grain-boundary width of 20 Angstroms (upper curve) and 2.4 Angstroms (lower curve). The image potential lowers the potential barrier quite significantly.

method (as outlined in Chapter 2) we obtain a value for gold of R at the Fermi level to be around 0.6, although the value depends extremely sensitively on the grain-boundary width.

From the M–S model, we have for the grain-boundary component of resistivity ratio:

$$\frac{\rho_0}{\rho} = 3\left(\frac{1}{3} - \frac{\alpha}{2} + \alpha^2 - \alpha^3 \ln(1 + \frac{1}{\alpha})\right) \quad \text{where} \quad \alpha = \frac{\lambda}{D_{50}} \frac{R}{1 - R}. \quad (3.15)$$

Due to the relationship between D_{50} and film thickness, we can calculate the resistivity of a thin film, as shown in Fig. 3.8. The trend is of the same form as that predicted by F–S theory.

Of course, the grain-boundary scattering is superimposed on surface scattering, so both must be considered to obtain a realistic model.

Assuming that the surface scattering (F–S term) and grain-boundary scattering (M–S term) terms can be described by relaxation times τ_{FS} and τ_{MS}, we can estimate the overall resistivity simply by calculating both terms separately and combining such that the total resistivity is described by a combined relaxation time:

$$\tau = (1/\tau_{FS} + 1/\tau_{MS})^{-1}. \quad (3.16)$$

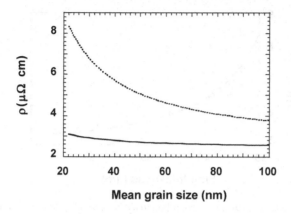

Fig. 3.8. Resistivity (grain-boundary component) versus film thickness (mean grain size) for gold films where $\lambda = 40$ nm; solid curve, $R = 0.1$, dotted curve, $R = 0.5$.

The effect of background (electron–phonon) scattering must then be considered separately, as in the presence of background and surface/grain boundary scattering, Mathiessen's rule (i.e. that the total resistivity of a sample is the sum of the temperature-dependent part and the part due to scattering from imperfections, which are independent) is no longer satisfied, so we can no longer combine relaxation times as above. As we are primarily interested in the *relative* importance of surface and grain-boundary scattering here, background scattering is of little consequence to our first-order analysis.

Within this framework, the net resistivity ratio of a thin film is found by remembering that resistivity is inversely proportional to mean free time, meaning we can combine the surface and grain-boundary components of resistivity simply by adding them together, i.e.

$$\rho = \rho_{FS} + \rho_{MS}. \tag{3.17}$$

This is shown in Fig. 3.9 using values for p of 0.008 and R of 0.23. Also shown is the calculation for $p = 0.6$ and R of 0.3, which predicts similar behaviour.

Therefore, it is actually not possible to experimentally determine either p or R of a thin film accurately. Traditionally, people have

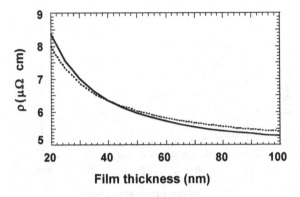

Fig. 3.9. Net resistivity (surface + grain boundary scattering) versus film thickness for a gold film where $\lambda = 40$ nm; solid curve, $R = 0.23$, $p = 0.008$, dotted curve, $R = 0.3$, $p = 0.6$.

assumed a value for one of these variables and used the measured resistivity to infer the other. However, there is a way around this which consists in looking at thin wires rather than thin films. In the absence of any dependence of the mean grain size on the wire width, the MS component will be a constant, depending only on wire thickness. The MS equation was arrived at by assuming a Gaussian distribution of grain sizes for mathematical simplicity. In reality, the grain size within metallic thin films tends to follow a lognormal distribution with a mean value denoted by D_{50} [15]. We can estimate the effective grain size distribution as a function of the linewidth by considering the fact that as the linewidth decreases below the film's D_{50}, the effective mean grain size will in fact depend on the linewidth. The average number of bamboo sections in a wire is given by

$$N_b = \int\limits_w^\infty f(D)\frac{D-w}{w}\,dD,$$

where

$$f(D) = \frac{1}{\sigma D\sqrt{2\pi}}e^{\left[-\left(\frac{1}{\sqrt{2}\sigma}\ln\left(\frac{D}{D_{50}}\right)\right)^2\right]}.$$

(3.18)

From which we determine that the average grain size (i.e. the average distance between grain boundaries) is given by

$$D_{\text{eff}} = \frac{D_{50}}{N_b + 1}.$$ (3.19)

Here, σ is the lognormal standard deviation of the grain diameters, D is the grain size and w is the linewidth. From this equation, we calculate that, on average, the mean distance between grain boundaries (D_{eff}) decreases as the wire width decreases, in the range $0.5D_{50} < w < 1.3D_{50}$, reaching plateaux above and below those limits. This is shown in Fig. 3.10, assuming typical values for $\sigma = 0.5$, and $D_{50} = 40$ nm.

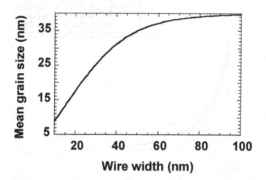

Fig. 3.10.   Dependence of mean grain size on linewidth for a gold film where $\sigma = 0.5$, and $D_{50} = 40$ nm. This is scalable for any ratio of linewidth/D_{50}.

Incorporating this into the MS model, i.e. if we replace D_{50} with D_{eff}, we find that the resistivity of a granular thin film increases faster with decreasing linewidth than we would have expected, as shown in Fig. 3.11.

The resistivity reaches a plateau at a wire width of about 60 nm or $1.5D_{50}$ as thereafter, the wire becomes polycrystalline rather than bamboo-like, and the average distance between grain boundaries is just D_{50}. When we combine the surface and grain-boundary components of resistivity, we end up with the trend shown in Fig. 3.12.

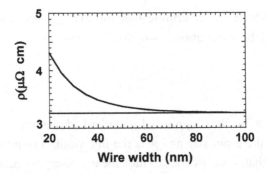

Fig. 3.11. Resistivity (grain-boundary component) versus linewidth for a gold film where $\lambda = 40$ nm, $D_{50} = 40$ nm, $R = 0.2$. The straight line shows the resistivity for a constant D_{50}.

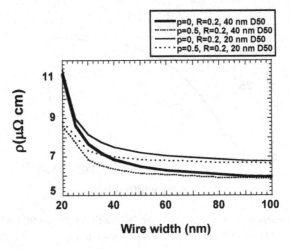

Fig. 3.12. Net resistivity versus linewidth for a gold nanowire where $\lambda = 40$ nm, for different values of D_{50}: 40 nm and 20 nm, and for fully diffuse and partially diffuse surface scattering.

What we can conclude from the calculation in Fig. 3.12 is that wires with a smaller mean grain diameter will tend to have (a) larger resistivity and (b) less prevalent surface scattering than those with larger grain diameters.

Since the late 1990s, several attempts have been made to combine surface and grain-boundary scattering models for comparison with experiment, with varying degrees of success [16–20]. As yet, there is no rigorous first-principles calculation combining these with background scattering.

3.5. Experimental Aspects: How to Measure the Resistance of a Thin Film

It has become standard practise when dealing with thin films to talk of something known as the "sheet resistance". For a bulk sample of cross-sectional area A and thickness w, the resistance $R = V/I = \rho w/A$. The sheet resistance is defined as $R_s = \rho/w$. For the case of a rectangular thin film of length x and width y, the resistance is $R = \rho x/wy$, which is just $R_s x/y$. When the sample is square therefore, the measured resistance is the same as the sheet resistance, which is also known as the *resistance per square*.

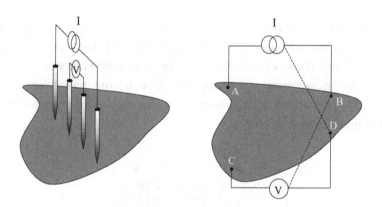

Fig. 3.13. (a) Four-point probe measurement; (b) van der Pauw technique — the resistance is measured as $(V_{CD}/I_{AB} + V_{CB}/I_{AD})/2$. V_{CD} is measured first and then two of the connections are swapped (B and D), at which point V_{CB} is measured.

Experimentally, resistance is measured using a four-point technique, as illustrated in Fig. 3.13. Clearly, the resistance given by the ratio of V/I is not the same as the sheet resistance, as the distribution of current within the sample will depend on both the sample and probe geometry. We therefore say that $R_s = CV/I$, where C is a geometric factor. When the sample has a well-defined geometry, it is possible to calculate this correction factor, but this is often not the case. This can be overcome using the van der Pauw method, as illustrated in Fig. 3.13, where the average of two four-point measurements is taken.

The situation is a little more complex in the case of devices, where we need to perform a four-terminal I/V measurement. This enables us to negate the contact resistance caused when attaching a small wire to a larger one. To see why this is so, we make use of the Büttiker formula [1] relating the current in terminal p to the electrochemical potential μ, at all other terminals, q, and the transmission probabilities T_{qp} and T_{pq} for electrons to travel from terminal p to q and vice-versa. The current I_p, is given by

$$I_p = \frac{2e}{h} \sum_q \left[T_{qp}\mu_p - T_{pq}\mu_q \right]. \tag{3.20}$$

If we apply this to a four-terminal configuration, as shown in Fig. 3.14, and measure the voltage across terminals 2 and 3, the resulting resistance is $R_{31}-R_{21}$, which gives the resistance of the wire section alone.

As two-terminal measurements give us the resistance of the device (R_d) plus all the interconnects to it (R_i), they can only sensibly be used in the case that $R_d \gg R_i$.

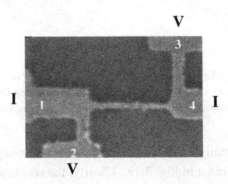

Fig. 3.14. Four-terminal configuration for measuring resistance.

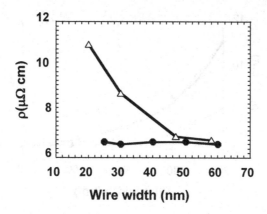

Fig. 3.15. Experimentally measured variation of resistivity on linewidth of gold nanowires; triangles and circles are for mean grain sizes of 40 nm and 20 nm, respectively [taken from Ref. 16].

A plot of the experimentally measured (in a four-terminal configuration) resistivity of gold nanowires as a function of the wire width is shown in Fig. 3.15 (open triangles) for wires with two different mean grain sizes (D_{50}), 20 nm and 40 nm.

We can see that the resistivity starts to increase once the wire width decreases below about 45–50 nm for the wires with larger grains, whereas no obvious size dependence is seen for the wires with smaller grains.

The most startling point to note here is that for the wires with smaller mean grain diameters, the resistivity appears to be independent of wire dimensions. This is a very strong indication that in practice, surface scattering is significantly less relevant than grain-boundary scattering in gold thin films and wires. In Fig. 3.16, we have re-plotted this along with fits to the above models.

These fits were made using the following parameters:

Wires with $D_{50} = 40$ nm:
Solid curve: $R = 0.54$, $\lambda = 40$ nm, fully specular scattering ($p = 1$)
Dotted curve: $R = 0.4$, $p = 0.5$, result scaled down by factor of 1.38.

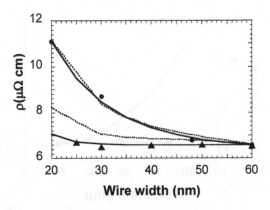

Fig. 3.16. Fits to experimentally measured variation of resistivity on linewidth of gold nanowires, from Fig. 3.15. Points (circles and triangles) are measured values.

Wires with $D_{50} = 20$ nm:

Solid curve: $R = 0.39$, $\lambda = 40$ nm, fully specular scattering ($p = 1$)

Dotted curve: $R = 0.4$, $p = 0.5$, result scaled down by factor of 1.38.

From the data, it is clear that the mfp is approximately 40 nm, so the bulk value of resistivity of 2.4 $\mu\Omega$ cm is correct. This means that the dotted curves do not represent reality as they predict a longer mfp than we observe. Again this means that at least for these types of sample, grain-boundary scattering is almost entirely responsible for the observed resistivity versus wire dimension that is experimentally observed. This is consistent with the fact that gold is known to exhibit a high degree of specularity as well as a high reflection coefficient. A true test of these models would therefore be to compare wires and films of different materials.

Whilst the agreement between theory and experiment is extremely good, we are implicitly assuming three points:

1. The wire surfaces are smooth and the cross-section of the wires is constant;

2. Grain boundaries are vertically oriented;

3. Our expression for mean grain size versus wire width is accurate.

In practice, neither of the first two will be the case, as the surfaces of e-beam fabricated structures tend to have a roughness of at least 1 nm (due to the granularity of the resist), and grain-boundaries will tend to be at a variety of angles to the surface normal. Variations in surface roughness will cause variations in the specularity parameter p, and the grain-boundary scattering will be sample-specific. Therefore, to obtain a more accurate range of values for p and R one should devise a differential resistance measurement, e.g. between a wire and a film or two wires to negate the effect of either grain boundary or surface scattering, respectively. The experimental data compels us to assume that our third assumption is justified, as otherwise we would expect the mean film grain size to have no bearing on the size-dependence of resistivity, whereas it in fact has a marked effect.

This leads us to the following conclusions regarding polycrystalline wires with dimensions comparable to the electronic mfp:

- When the wire width is comparable to the mean film grain size, grain-boundary scattering is the dominant source of increased resistivity.
- When the wire width is below approximately 0.5 times the mean film grain size, surface scattering will become important and approaches the same order of magnitude as grain-boundary scattering as the width decreases.

We can see that the techniques for measuring the resistance of nanostructures and for interpreting those measurements are indeed quite mature although much remains to be understood from a theoretical standpoint. In particular, it is only possible to calculate the effect of defects and phonons (background scattering) on transport through *small* devices, i.e. ones with only a few defects. In reality, even the smallest structures which we can fabricate will have defects, particularly metallic structures.

References for Chapter 3

1. S. Datta, *Electronic Transport in Mesoscopic Systems* (Cambridge University Press, Cambridge, 1995).
2. N. W. Ashcroft and N. D. Mermin, *Solid State Physics* (Saunders College Publishing, Philadelphia, 1976).
3. C. Kittel, *Introduction to Solid State Physics*, 6th edn. (Wiley, 1986).
4. K. Fuchs, *Proc. Camb. Phil. Soc.* **34**, 100 (1938).
5. E. H. Sondheimer, *Adv. Phys.* **1**, 1 (1952).
6. D. K. C. MacDonald and K. Sarginson, *Proc. Roy. Soc. A* **203**, 223 (1950).
7. R. B. Dingle, *Proc. Roy. Soc. A* **201**, 545 (1950).
8. R. G. Chambers, *Proc. Roy. Soc. A* **202**, 375 (1950).
9. A. F. Mayadas, *J. Appl. Phys.* **39**, 4241 (1968).
10. A. F. Mayadas, M. Shatzkes and M. Janak, *Appl. Phys. Lett.* **14**, 345 (1969).
11. A. F. Mayadas and M. Shatzkes, *Phys. Rev. B* **1**, 1382 (1970).
12. M. A. Schneider, M. Wenderoth, A. J. Heinrich, M. A. Rosentretter and R. G. Ulbrich, *J. Elect. Mat.* **26**, 383 (1997).
13. C. Durkan, M. A. Schneider and M. E. Welland, *J. Appl. Phys.* **86**, 1280 (1999).
14. C. Durkan and M. E. Welland, *Ultramicroscopy* **82**, 125 (2000).
15. Y.-C. Joo and C. V. Thompson, *J. Appl. Phys.* **76**, 7339 (1994).
16. C. Durkan and M. E. Welland, *Phys. Rev. B* **61**, 14215 (2000).
17. W. Steinhogel, G. Schindler, G. Steinlesberger and M. Engelhardt, *J. Appl. Phys.* **97**, 23706 (2005).
18. W. Steinhogel, G. Schindler, G. Steinlesberger, M. Traving and M. Engelhardt, *Phys. Rev. B* **66**, 75414 (2002).
19. H. Marom and M. Eizenberg, *J. Appl. Phys.* **96**, 3319 (2004).
20. W. Wu, S. H. Brongersma, M. van Hove and K. Maex, *Appl. Phys. Lett.* **84**, 2838 (2004).

Chapter 4

Scanning-Probe Multimeters

We saw in the previous chapter that transport measurements on macro- and mesoscopic devices are performed using probes, preferably at least four of them. It is generally assumed that the presence of the probes does not in any way perturb the device being measured. This is no longer the case once the device has dimensions at the nanoscale [1–3]. In this chapter we will look at the techniques used to locally probe currents and voltages in nanostructures.

Since the advent of the scanning tunnelling microscope (STM) in 1982 [4], a whole new world at the nanoscale has become accessible. It is now possible to image and even manipulate surfaces at the molecular and atomic level [5–9]. However, STM as a technique has the drawback that it is only applicable to conducting surfaces. As artificial current-carrying mesostructures (i.e. structures with dimensions from 100–1000 nm) necessarily consist of a conductor on top of an insulator, STM is of no practical use here, except for instance for measurements of the conductance of single molecules and as we shall see later, for potentiometric studies of conducting surfaces. It is here that the flexibility of Atomic Force Microscopy (AFM) can be exploited.

4.1. Scanning-Probe Microscopy: An Introduction

Scanning probe microscopy, having been invented in the early 1980s, has flourished as a research field. There are now many different

forms of SPM, and a large number of companies producing microscopes
and related consumables, which broadly fall into three classifications:

1. STM
2. AFM
3. SNOM (Scanning near-field optical microscopy)

There are now several texts which describe the operation of each of
these in detail [10–12], so here we will just consider the general aspects
of SPM which are related to transport measurements.

The basic principle behind the operation of all SPMs is that a
sharp probe tip is scanned in close proximity to a sample surface, whilst
some interaction between the two is used to generate an image of the
surface.   Central to this is a feedback loop which controls the tip–
sample distance during scanning in order to maintain a constant
interaction, as shown in Fig. 4.1.   Different types of SPM utilize
different interactions.

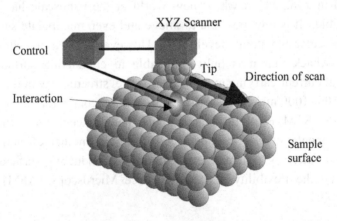

Fig. 4.1. Overview of SPM operating principle.

As we are interested in electrical properties of nanostructures,
we will now briefly consider the basis of general STM and AFM

imaging and measurement (SNOM is not relevant in this context), before considering specific measurement modes.

4.2. Scanning Tunnelling Microscopy

4.2.1. *Basic principles*

We have already encountered the concept of electron tunnelling in Chapter 2, where we looked at the case of 1D tunnelling in a device structure, where we implicitly assumed the materials on both sides of the barrier acted as electron reservoirs with an endless supply of electrons. When we then looked at field emission, we assumed that the electrons were tunnelling out of a conductor into free space. The principle of the STM is that we tunnel from a tip into a sample (or vice-versa, depending on the relative polarity of tip and sample) via free space, which unlike a 1D nanowire has an infinite number of possible electron modes. This has the effect that the number of electron modes, or more correctly the density of states of the tip and sample now plays a crucial role in determining the characteristics of the current flow. The current is effectively given by:

$$I \propto \int_{eV}^{0} \mathcal{D}_s \mathcal{D}_t T dE, \qquad (4.1)$$

where $\mathcal{D}_s$ and $\mathcal{D}_t$ are the density of states of the sample and tip, respectively, and T is the transmission probability of the barrier, as calculated in Chapter 2. Nowadays, quite sophisticated calculations can be performed which give extremely accurate representations of STM images. The analysis which we have used so far is certainly useful for illustrative purposes, but presupposes that the electron wave-functions are 1D plane waves. The most commonly used general approach is that of Bardeen [13], which is based on time-dependant perturbation theory. In this approach, the tip and sample are treated as independent systems within which we determine the electron wave-functions by solving

Schrödinger's equation (taking into account the material and geometry of tip and sample). Once these wave-functions are known, time-dependant perturbation theory is used to determine the rate of electron transfer from one electrode to another, the key assumption of course being that the interaction between tip and sample is very weak. This rate will depend on the amount of overlap between the wave-functions within both electrodes, which is given by the tunnelling matrix element, M:

$$M = \frac{\hbar}{2e} \oint_S \left(\psi_t^* \frac{\partial \psi_s}{\partial z} - \psi_s \frac{\partial \psi_t^*}{\partial z} \right) dS, \qquad (4.2)$$

where ψ_s and ψ_t are the wave-functions within the sample and tip, respectively, and the integral is over a separation surface between the tip and sample. This is illustrated in Fig. 4.2.

Using Fermi's golden rule, the probability that an electron will tunnel from one electrode into another, assuming perfectly elastic tunnelling is

$$\frac{2\pi}{\hbar} |M|^2 \, \delta \left(E_t - E_s \right). \qquad (4.3)$$

Summing over all possible transitions, the total tunnelling current is found to be

$$I = \frac{4\pi e}{\hbar} \int_{\mu_s}^{\mu_t} \mathcal{D}_t(E)\left(E_F + \varepsilon \right) \mathcal{D}_s \left(E_F - eV + \varepsilon \right) |M|^2 \left[f\left(E_F - eV + \varepsilon \right) - f\left(E_F + \varepsilon \right) \right] d\varepsilon.$$

$$(4.4)$$

If we compare this to the expression we calculated in Chapter 2 for transport through any device,

$$J_{\text{total}} = \frac{2e}{\hbar} \left(\int_{\mu_r}^{\mu_l} \mathcal{D}(E)(f_l(E) - f_r(E_r))T(E)dE \right). \qquad (4.5)$$

We see that the two are *almost* identical, except that $T(E)$ and $\mathcal{D}(E)$ have been replaced by $|M|^2$, and $\mathcal{D}_t\mathcal{D}_s$, respectively. As mentioned earlier, this difference is due to the fact that when tunnelling through a vacuum barrier, the current now depends on the number of available states *on both sides of the barrier*, and the two electrodes are effectively probing each other.

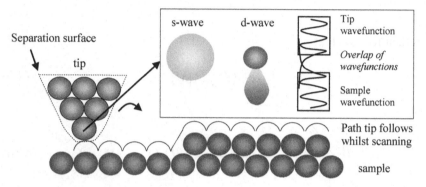

Fig. 4.2. Principle of operation of STM. Inset shows two possible tip states, i.e. where the apex atom has its outer electrons in *s* or *d* orbitals. On the right, we have indicated the overlap between tip and sample wave-functions, which gives rise to tunnelling.

The above expression for the current can be further simplified by recognizing that experiments are typically performed at or below room temperature and at low bias, with the consequence that the Fermi function is step-like and M is essentially constant, giving the following:

$$I = |M|^2 \int_0^{eV} \mathcal{D}_t\left(E_F + \varepsilon\right)\mathcal{D}_s\left(E_F - eV + \varepsilon\right)d\varepsilon. \qquad (4.6)$$

If we choose a tip with a DoS which is constant over the energy range determined by eV, then we find that

$$I = |M|^2 \mathcal{D}_t \int_0^{eV} \mathcal{D}_s\left(E_F - eV + \varepsilon\right)d\varepsilon, \qquad (4.7)$$

$$\frac{dI}{dV} = |M|^2 \mathcal{D}_t \mathcal{D}_s \left(E_F - eV + \varepsilon \right). \tag{4.8}$$

Now, if we make use of the fact that

$$\frac{1}{eV} \int_0^{eV} \mathcal{D}_s \left(E_F - eV + \varepsilon \right) d\varepsilon = \langle \mathcal{D}_s \rangle \text{ i.e. the average value of } \mathcal{D}_s.$$

It then follows that

$$I = eV \langle \mathcal{D}_s \rangle \mathcal{D}_t |M|^2 \tag{4.9}$$

so

$$\frac{dI}{dV} \Bigg/ \frac{I}{V} = \frac{\mathcal{D}_s \left(E_F - eV + \varepsilon \right)}{\langle \mathcal{D}_s \rangle} \propto \mathcal{D}_s \left(E_F - eV + \varepsilon \right). \tag{4.10}$$

There are a number of consequences of this for STM operation:

- Whilst the tip is scanning over the surface maintaining a constant current level, the resulting "topographic" image is nothing other than a spatial map of the density of states, so it is strictly not correct to consider STM images as representing the topography of a surface.
- The current depends on $|M|^2$, which is essentially the same as $T(E)$. We saw in Chapter 2 that the transmission probability for a thin tunnel barrier of height φ and width z scales as

$$T \propto e^{-\frac{2z\sqrt{2m\varphi}}{\hbar}} \tag{4.11}$$

so the tunnel current depends exponentially on the width of the tunnel gap. Inserting a typical value for φ of 4 eV, we find that if the tip–sample distance changes by as little as 0.1 nm, T changes by almost an order of magnitude. This fact is responsible for the extremely high resolution of STM, but also means that these microscopes must be mechanically robust and stable and isolated from external vibrations.

• Early attempts at estimating matrix elements used the assumption that the tip wave-function was spherically symmetric, i.e. tip states are *s*-type wave-functions. This approach, developed by Tersoff and Hamann [14] was useful in explaining many aspects of STM operation, but could not satisfactorily explain why the lateral resolution is as high as it is. It is often the case that the tip atom is in a *d*- or *f*-type state, the orbitals of which are much more confined than *s* orbitals. See Fig. 4.2.

• If one sweeps the voltage whilst measuring the current, calculates the derivative and normalizes that to the junction conductance (I/V), it is possible to measure the (unscaled) density of states of the sample. Alternatively, a small ac voltage at frequency ω can be added to the dc sample or tip bias, and lock-in detection of the tunnel current at the frequency ω will give a similar result. This is known as *tunnelling spectroscopy* which has become an extremely powerful and useful analytical technique, and has been used to study all manner of systems, from conductors [15,16], semiconductors [17–22], superconductors [23–26] and molecules [27–31]. Magnetic tips, i.e. ones with a spin-dependant density of states have been used to map out the magnetization of surfaces with atomic resolution, in a technique known as spin-polarised STM [32–36].

To estimate the lateral resolution of STM, we make use of the fact that the tunnel current depends exponentially on the tip–sample distance. If the STM tip has a radius of R nm, and the minimum tip–sample distance is d, then the distance between the tip surface and the sample varies with x as $z = d + x^2/2R$. If we define the resolution as the lateral distance at which the current drops to around 10% of its maximum value, we get a value for x of around $(0.1/2R)^{0.5}$. This means that in order to achieve atomic resolution (i.e. $x \sim 0.1$ nm) the tip radius should be no larger than 5 nm. In actual fact, the tip radius is often much larger than this, and atomic resolution is still achieved, indicating that there must be an asperity, possibly even a single atom at the end of most STM tips. In many cases, this single atom can come from the sample itself. The nature of this apex atom will determine the resolution, image contrast and spectroscopy data obtained in any given STM experiment, so extensive cleaning is often carried out on STM tips in order to improve repeatability.

In all cases, there will be a small amount of inelastic tunnelling, e.g. phonon-assisted tunnelling, a threshold process which manifests itself as a change in slope in the *I–V* characteristic, as shown in Fig. 4.3. The exact energy of the phonon modes can be found simply by looking at the second derivative of *I*. This technique has been used recently to probe vibration modes in single molecules [37–40].

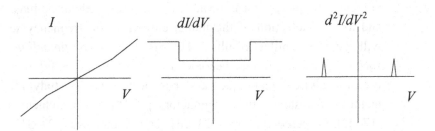

Fig. 4.3. Effect of inelastic tunnelling serves to increase tunnel current, as additional transmission channels become available. Vibrational modes of the sample are seen as peaks in d^2I/dV^2 spectra.

4.2.2. Scanning tunnelling microscopy in practise

STM has been demonstrated time and time again to be capable of atomic resolution on surfaces. However, apart from the case of HOPG (graphite), this comes at a price — UHV (Ultra-High Vacuum). UHV conditions (i.e. base pressure of 10^{-10} mbar or lower) are required in order to maintain a clean STM tip and sample for a number of hours. Under ambient conditions, all surfaces are coated with several monolayers of adsorbed "contamination" — mostly water and hydrocarbons. In order to probe the underlying surface, samples need

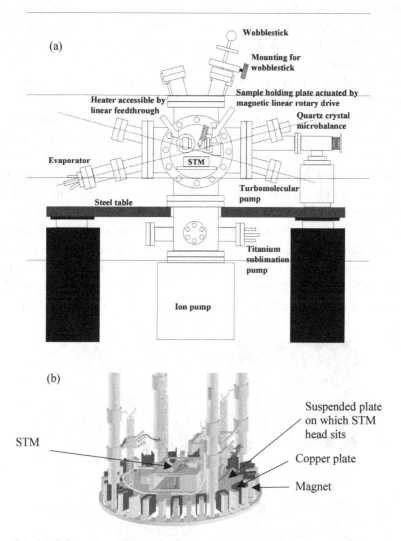

Fig. 4.4. (a) Schematic of the author's UHV STM system. Samples and tips are introduced into the UHV chamber via a fast-entry lock, and are inserted into the heater/STM head using a wobblestick manipulator. External vibration isolation is achieved by placing the STM on air legs which are sitting on separate foundations; (b) internal vibration isolation is managed using spring suspension with eddy-current damping.

to be cleaned, and the amount of time for which they remain clean will depend on the pressure of their local environment. Under ambient conditions, any surface will remain clean for only a fraction of a second, and under deep UHV conditions ($\sim 10^{-11}$ mbar) this can be extended to several days. In a typical STM experiment, the tip (typically etched W or mechanically cut Pt/Ir tips are used) and sample are cleaned immediately prior to imaging, usually by repeated cycles of thermal annealing and ion sputtering. The author's own STM is illustrated in Fig. 4.4 to give an impression of the appearance of a typical system. Vibration isolation is paramount, as the floor in a typical building will be vibrating at several Hz, with an amplitude of several μm. These vibrations must be damped to around 1 pm or below for the STM to work effectively. There are many different ways of achieving this level of isolation: (i) the STM head can be made small and stiff with natural resonance frequencies in the kHz range, or (ii) the STM can be mounted on vibration isolators. For ease of use, many researchers have gone for the second option, as small STMs can be rather difficult to use, especially under UHV conditions. In Fig. 4.5, there are a series of STM images of different surfaces demonstrating the power of this technique.

Fig. 4.5. STM images of (a) Si(111), (b) HOPG, (c) Au(111), (d) monatomic steps and pits on Au(111) surface.

STM is mostly used to map out surfaces and perform localised transport measurements on them, from which information about the density of states may be extracted.

Scanning tunnelling spectroscopy, as has already been mentioned, is an extremely powerful extension to the capabilities of STM. One can measure (in arbitrary units) the density of states of surfaces and adsorbates, as illustrated in Fig. 4.6 [41–43]. For semiconductors, one can measure the band-gap, for metals, one can investigate surface and image states, and for molecules/adsorbates, one can measure transport and spatially map the density of states.

If we remember from Chapter 1 that the current, I for small voltages, V is

$$I \propto V e^{-\frac{2s\sqrt{2m\phi}}{\hbar}}, \tag{4.12}$$

$$\Rightarrow \text{for constant voltage, } \phi = -\frac{\hbar^2}{8m}\left(\frac{\partial \ln I}{\partial s}\right)^2. \tag{4.13}$$

This quantity is often referred to as the *effective barrier height*. It may be measured by modulating the gap width, s (by a very small amount, typically < 0.1 Å), and measuring the corresponding modulation in the current. This is a useful quantity to measure, as changes in the local barrier height are due to changes in surface structure, be that a reconstruction or an adsorbate.

There has been some controversy over the dependence of the barrier height versus distance, as a simple calculation such as the one we presented in Fig. 2.44 indicates that the barrier height will decrease once the gap width decreases below around 0.5–0.6 nm. However, a key assumption in arriving at Eq. (4.13) is that the voltage is a constant. This is certainly not the case in reality, as the voltage is applied between the sample and a virtual earth on the tip (or vice-versa) which is connected to a current pre-amplifier of finite input impedance. A typical amplifier used in STM experiments will have an input impedance of ~ 100 kΩ, and when we consider that the tunnel resistance can be as low as a few 100 Ω at contact, this becomes an important issue. In effect,

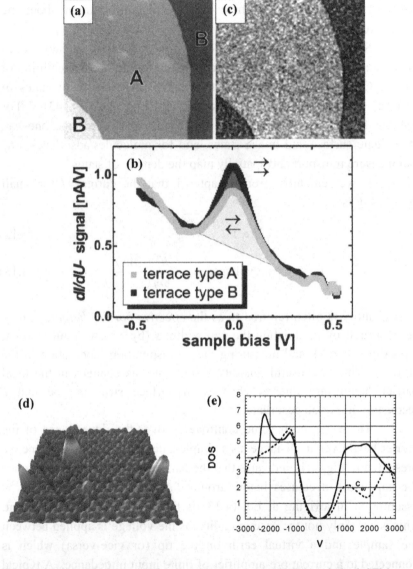

Fig. 4.6. Examples of tunnelling spectroscopy. (a)–(c) Spin-polarised STM on a magnetic sample [42]. In (c), arrows represent relative magnetic orientation of tip and sample. (d) STM image of C_{60} molecules on Si(111), (e) density of states of Si(111) and a single C_{60} molecule.

once the tip–sample distance decreases below a certain value, the voltage between them will decrease, and the effective barrier height will appear to decrease, unless the voltage across the tunnel junction is measured directly. This has been looked at by Besenbacher [44] in the mid 90s, who saw that the voltage does indeed decrease, and by taking that into account, the effective barrier height as measured by STM is found to be independent of tip–sample distance. Why is this the case? Our simple model indicates that the barrier height should decrease due to the image potential. It turns out that this does indeed happen, but it is counter-balanced by the forces between the tip and sample. It has long been known that there can be a significant force (10^{-9} N and higher) between an STM tip and a sample surface during scanning, and indeed this is the accepted reason why many layered materials (e.g. graphite) appear to have anomously large atomic corrugations in STM images (i.e. ≥ 1 Å) [45]. In terms of explaining the observations on barrier heights, as the tip is approached to the sample, when the gap is ≤ 2 Å, attractive forces pull the tip towards the sample, so the gap width starts decreasing faster than expected, and the current starts increasing at a greater rate, with the net result that the barrier appears to be getting lower. We must also consider that whilst we have shown that the barrier height should decrease with decreasing gap width, it is more complicated in reality, due to the 3D nature of real STM tips. A hole does open up in the potential barrier directly below the tip as indicated in Fig. 4.7, but this is partially balanced by the fact that a constriction on the electrons (this is like a potential well) in the lateral plane reduces their momentum along the surface normal, reducing the current with the result that an effective barrier still appears to be present [46,47].

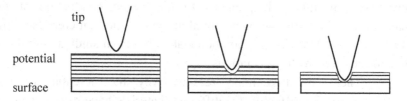

Fig. 4.7. Potential barrier around an STM tip–sample junction. As the tip gets closer to the sample, the image force causes a hole to open in the barrier at the tip apex.

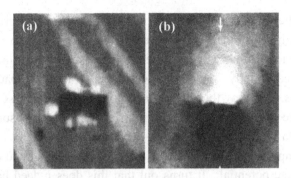

Fig. 4.8.   61 nm × 75 nm area of 3 nm thick Bi film with a 2.4 nm deep hole. (a) STM topography, (b) potential.

The fact that the tip–sample interaction is large enough to have such a marked effect means that STM image interpretation is even more complex than was first envisaged, and one must be particularly careful when measuring the height of adsorbed structures on surfaces.

STM can also be used to measure surface potential — to see how, one need only remember that the magnitude of the tunnel current depends on the potential difference between the tip and the sample. This can be re-stated as follows: for a fixed tip–sample separation, local variations in surface electrochemical potential will introduce variations in the tunnel current. Therefore, to measure the surface potential, a second feedback loop is used to vary the tip-sample bias until zero tunnel current is obtained at each image point, whilst holding the tip–sample distance constant. This technique is called "scanning tunnelling potentiometry" [48–51]. In this way, a resolution of surface potential of the order a few microvolts has been demonstrated. Coupled with the inherent lateral spatial resolution of STM, this technique has been immensely useful in helping us to understand conduction at the nanoscale. The variation of potential around single scatterers (see Fig. 4.8) and the reflectivity of grain boundaries has been studied, and found to correlate well with expectations.

There are two main reasons why the potential of two surfaces might be different: (a) different materials have different Fermi energies and work-functions, and (b) there can be a thermovoltage. A thermovoltage arises whenever there is a temperature difference between two materials. In the context of STM experiments, a thermovoltage arises between an STM tip and sample whenever both are at a different temperature. If one holds a sample at a constant temperature whilst heating the tip, there will be a shift in the electron distribution of the tip around the Fermi level, and this will result in a net tunnelling current and voltage difference between tip and sample, even under zero applied bias. The actual thermovoltage will depend very sensitively on the density of states of both the tip and sample [52,53], so by measuring it, one may obtain information about the density of states of a sample with greater accuracy than can be obtained by tunnelling spectroscopy alone. Thermovoltage in an STM is measured in exactly the same way as surface potential (mentioned above), the only other difference being that the tip or sample must be heated to generate a measurable temperature difference, usually by a laser. Some example images are shown in Fig. 4.9.

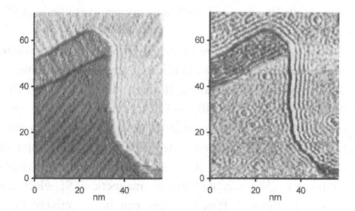

Fig. 4.9. Thermovoltage imaging with STM, on Au(111) surface. Left image: STM image of topography; right image: thermovoltage (colour scale corresponds to a range of ~ 70 μV thermovoltage). Reprinted from Ref. 52.

Whilst STM has been shown here to be a powerful analytical tool, it cannot be used to study insulating samples (the exception being monolayers of insulator on top of a conductor). As has been mentioned earlier in this book, we are interested in looking at devices which typically consist of a conductor/semiconductor grown on top of an insulator. STM therefore cannot be used to study such samples, so we need to resort to AFM-based techniques, which we will now consider.

4.3. Atomic Force Microscopy

In atomic force microscopy [54,55], which was developed a few years after the STM, the interaction between tip and sample is based on forces rather than on currents as is the case in the STM. These forces are usually sensed by placing the tip on a micro-machined cantilever beam (these are now mass-produced and may be purchased for a few dollars), the deflection of which is measured using a position-sensitive detector. The simplest way to implement this is to use the *"beam deflection"* method, whereby a laser is focused on the cantilever near the tip end, and as the cantilever bends, the reflected laser beam will move. This motion can easily be detected with a quadrant photodiode (QPD), as illustrated in Fig. 4.10.

In this way and using an off-the-shelf laser and photodiode, it is possible to detect cantilever deflections of the order 0.01 nm. Even better sensitivity can be achieved through the use of tunnelling or interferometry [56,57]. Whilst STM is clearly a very powerful technique, AFM is far more so, due to the generality of its operating principle. AFM has also demonstrated atomic resolution capability in recent years. There exist many different types of force at the nanoscale, including friction [58], van der Waals, magnetic [59], electrical and thermal, among others. Different forces can be selectively measured simply by choosing a suitable tip, e.g. if one wants to measure the topography of a surface, a standard Si or Si_3N_4 tip should be used; magnetic fields on a sample surface can be measured by depositing a thin magnetic film on the tip, friction can be measured by looking at how a cantilever beam twists during scanning, and, most relevant for this text, electrical forces can be measured in a variety of ways, which we will look at shortly.

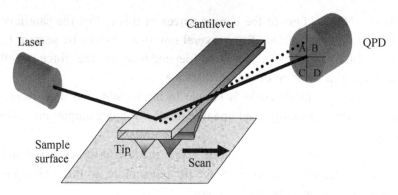

Fig. 4.10. Schematic set-up of AFM. Solid and dotted curves show laser beam path for two different cantilever deflections. Photodiode is split into four separate quadrants, A–D. The cantilever deflection is determined by measuring the laser intensity on quadrants $A + B - (C + D)$.

First, we will briefly consider the main modes of operation of AFM before considering how this relates to electrical measurements.

4.3.1. *Modes of operation of AFM*

There are three primary modes of operation of AFM in use today, as well as many derivatives and extensions of these. In chronological order when these techniques were developed, they are:

1. Contact mode;
2. Non-contact mode;
3. Intermittent contact mode (otherwise known as Tapping mode®).

Contact mode

In contact mode, as the name suggests, the tip is placed in intimate contact with the sample. During scanning, the normal force is typically maintained at a few nN (10^{-9} Newtons), which, when spread over a contact area of a few nm^2, corresponds to a local pressure of the order 10^9 Pa. In order to minimise these forces, the cantilevers used for contact mode AFM tend to have fairly low stiffness, of the order

0.01–0.5 Nm^{-1}. One of the consequences of this is that the cantilevers can snap onto the surface from several nm away, as can be seen in Fig. 4.11, which shows the measured dependence of the force on the cantilever as a function of sample distance.

Contact mode AFM is simple to implement, and until recently was the most commonly used imaging mode. Some examples are shown in Fig. 4.12.

The drawback of this mode is that the lateral forces can be quite substantial, leading to damage of the tip and sample, so this technique is most useful for imaging hard samples.

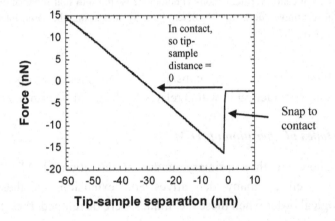

Fig. 4.11. Force–distance measurement in contact mode AFM.

Fig. 4.12. Contact mode AFM images of (a) Au nanowire, image size = 1 μm × 450 nm; (b) Mo oxide nanocrystal on Mo foil, image size = 2.5 μm × 2.5 μm.

Non-contact mode

Shortly after the development of AFM, another method was introduced, this time using the resonant behaviour of AFM cantilevers to increase the force sensitivity. In the most common implementation of this technique, the cantilever is mounted on a small piezoelectric plate and oscillated by a few nm at or near its fundamental resonance frequency, and scanned at a height (a few nm) where it is sensitive to the attractive van der Waals forces, which extend to around 1 nm from the sample surface (Fig. 1.15). In this way, the tip does not actually touch the surface, and lateral forces are negligible. As the tip approaches the sample, its oscillation amplitude will decrease, the phase will change, and the resonance frequency of the cantilever will shift (to a lower value), as shown by Eq. (4.14):

$$\Delta f \propto \frac{\partial F}{\partial z},\tag{4.14}$$

i.e. the resonance frequency shifts due to the force gradient.

An image may be obtained by using any of these signals for feedback control, the simplest being the amplitude signal (a simple rectifier can be used to obtain this signal, or a lock-in amplifier can be used to obtain this and the phase signal), and the most sensitive being the frequency shift, for which a phase-locked loop is needed (this is known as FMAFM, or frequency modulation AFM).

In order to obtain the highest resolution, the tip needs to be close enough to the surface for the force gradient to be high enough to produce a measurable frequency shift, which typically occurs at a distance of around 1 nm. Unfortunately, under ambient conditions, all surfaces have a layer of contamination consisting of water and hydrocarbons, which is generally a few nm thick. In contact mode AFM, the tip is immersed in this layer, so it does not matter much, but in non-contact AFM, it effectively means that the tip cannot get close enough to the true surface to image it properly, and any attempt to do so would result in the tip sticking to the contamination layer. However, under UHV conditions where the tip and sample are atomically clean, true atomic resolution has

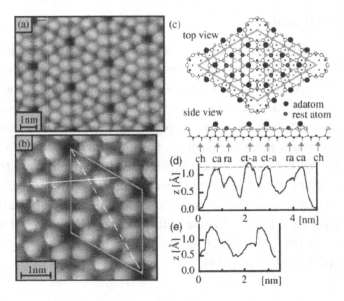

Fig. 4.13. Non-contact mode AFM images of Si(111) 7 × 7 surface. Images were generated by measuring the frequency shift of the AFM cantilever [60].

been demonstrated a number of times with an even higher resolution than can be obtained using STM [Ref. 60 and references therein], as shown in Fig. 4.13.

Intermittent contact (tapping) mode

The most recently developed mode of AFM operation is essentially a cross between contact and non-contact mode. As in non-contact mode, the cantilever is oscillated at or near its resonance frequency. The difference is that the oscillation amplitude is rather larger, at several tens of nm. This means the cantilever has enough energy to prevent the tip snapping onto the surface or getting trapped in the contamination layer. It also means that the tip "samples" the entire force gradient all the way from far away to contact, so the force changes from attractive to repulsive, making quantitative measurements of the force impossible. Distance regulation and hence imaging is typically done by maintaining constant oscillation amplitude, usually at around

90–95% of the free amplitude (i.e. the amplitude when the tip is far away from the sample). As an imaging technique, this is perhaps the most commonly used one now, as it has a very good resolution at the same time as having very low lateral forces on the sample. This is the technique of choice for looking at soft matter. In standard operation, the phase of the cantilever oscillation is monitored at the same time as the amplitude. Phase images, whilst again extremely difficult to quantify, reveal material differences. This is useful when imaging samples, particularly copolymers or self-assembled monolayers with little or no measurable topographic variation, but with distinctly different materials, as illustrated in Fig. 4.14. Tapping mode can also be used under liquid environments, although the reduced Q-factor of the cantilever due to hydrodynamic damping reduces the sensitivity.

Each of these three modes of operation may be employed to measure the potential of a surface, although slightly different information is obtained in each case, as we will see shortly.

We will now turn our attention to the various AFM-based techniques which may be employed for the electrical characterisation of nanostructures.

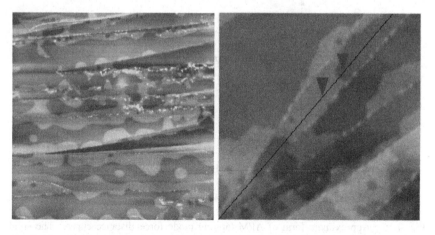

Fig. 4.14. Tapping mode AFM images (a) 7 μm × 7 μm and (b) 1.5 μm × 1.5 μm images of BDPA molecular islands on HOPG. These and many other materials are too soft to be imaged using contact mode AFM.

4.3.2. *Kelvin-probe force microscopy*

This is an extension of non-contact AFM, based on the fact that when a voltage (V_0) is applied between a conducting AFM tip and a sample, there is an attractive electrostatic force given by:

$$F = \frac{V_0^2}{2} \frac{\partial C}{\partial z},$$
(4.15)

where C and z are the capacitance and distance between the tip and sample, respectively. The exact value of C will depend on the tip's geometry and the sample's conductivity. This is a long range force, which may be used to detect local variations in surface potential in noncontact mode AFM, or even in "lift-mode" AFM, where the tip is scanned at a fixed height above the surface with the feedback loop turned off, and any electric field due to potential variations in the surface will cause the cantilever to deflect as it scans over the surface. Figure 4.15 shows the difference between approach curves with and without a voltage applied.

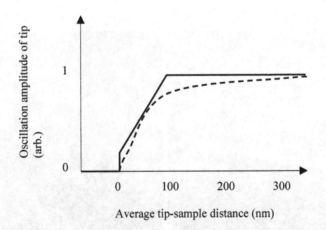

Fig. 4.15. Approximate form of AFM tapping mode force–distance curve. The solid and dotted curves are without and with a dc voltage applied to the sample, respectively. In this case, the peak–peak free oscillation amplitude of the cantilever beam is around 100 nm.

and also the imaging of trapped charges [65]. This technique has also been investigated as a means of high density data storage, as one may generate trapped surface charge with a conducting tip and subsequently read out using the tip in KPFM mode [66]. High-frequency probing of working devices has even been demonstrated by using V_0 and V_s at slightly different frequencies (ω_0 and ω_s), and by monitoring the tip deflection at the beat frequency, $\omega_0 - \omega_s$ [67,68].

4.3.3. Conducting mode AFM

Whilst substantial theoretical and experimental progress has been made with KPFM during recent years, the fact remains that the correlation between measured and actual potential is not yet fully understood for complex samples, and the resolution has not surpassed ~ 20 nm without resorting to exotic tip geometries. For this reason, much interest has arisen in conducting mode AFM, (C-AFM) [69]. This is a technique enabling one to directly measure surface potentials, with the superior resolution of contact or intermittent contact (*tapping*) mode AFM. The principle of operation is shown in Fig. 4.17. The tip is grounded through a limiting resistor, and the contact potential of the surface can be determined by measuring the current flowing through the tip.

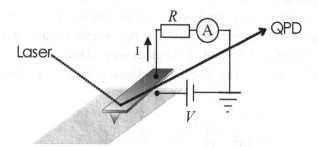

Fig. 4.17. Principle of conducting AFM. A voltage, V is applied to the sample, and a current, I flows through the tip. This current is limited by resistor, R, and is measured with ammeter, A. R is chosen to be much larger than the tip–sample resistance, so from the relationship $V = IR$, the surface potential may be measured. QPD is the quadrant photodiode used to detect the cantilever deflection.

In contact mode when a voltage is applied to the sample, the tip current as a function of position on the sample surface (r) is given by:

$$I(r) = \frac{V}{R(r)} + C(r)\frac{\partial V(t)}{\partial t} + V\frac{\partial C(r)}{\partial t} + \alpha(r)AV^2 e^{\frac{-\beta(r)}{V}} + \alpha V e^{-\gamma(r)z},$$

$$(4.20)$$

where R is the contact resistance, C is the tip–sample capacitance, and the other constants are related to either material or physical constants. The first term is the Ohmic contribution, the second and third are induced currents due to time variation of the fields around the tip, and the last two terms are tunnelling contributions. Depending on the nature of the contact area, the various terms will differ in importance.

When the tip is contacting a conductor, the Ohmic term will dominate, whereas when the tip is contacting an oxide layer, the Fowler–Nordheim [70] term (the second-last term above) dominates and finally, when there are time — varying fields present, the inductive terms dominate. Fowler–Nordheim tunnelling is essentially field emission, which occurs once the applied bias exceeds the tunnel barrier height. Meaningful measurements require a stable contact whereby we want to be able to ascribe differences in the tip current purely to differences in the surface potential. We can see from above that variations in the contact resistance will also produce differences in the measured current. Changes in the contact resistance will be due to changes in both the resistivity of the sample and also the tip–sample contact area. In the simplest model due to Hertz [71] describing elastic contact between a sphere and a plane in the absence of adhesive forces, the contact area is given by

$$A = \pi a^2 = \pi \left(\frac{3Fr}{4} \cdot \left(\frac{1-v_1^2}{E_1} + \frac{1-v_2^2}{E_2} \right) \right)^{2/3}, \qquad (4.21)$$

where F is the applied load, r is the tip radius of curvature, a is the contact radius, and $v_{1,2}$, $E_{1,2}$ are Poisson's ratio and Young's moduli for the tip and sample, respectively. For a gold tip on a gold sample at

50 nN applied load, and for a radius of curvature of the tip of 50 nm, the contact area is of the order $16*10^{-18}$ m^2. Thus, the radius of the contact is of the order 2 nm, which is well below the mfp, meaning we can confidently apply Sharvin's formula to calculate the contact resistance. It should be noted at this point that the contact area as calculated in this way will be lower than the actual value as adhesive forces will be present and plastic deformation of the tip and sample will occur, both of which will tend to increase the contact area.

One of the first uses of C-AFM was in the area of spreading resistance profiling (SRP). This is a technique whereby carrier concentration profiles in Si are measured by means of a point contact. This is typically achieved using a conducting indenter which presses into the Si surface until a stable electrical contact is formed. I/V measurements then yield a value for the local resistivity through the spreading resistance, from which carrier concentrations may be inferred.

The main difficulty with this technique lies in the fact that the pressures involved (of order 10 GPa) locally distort the crystal structure and hence the carrier concentration and cause band bending and band-edge shifting. Also, the contact size is typically in the 100 μm range, making high resolution spatial maps of carrier concentration unattainable. For this reason, work has been done using C-AFM tips as nano-indenters for nano-SRP measurements [72,73]. The advantages now are that the loads are significantly reduced to be of the order 200 μN, with contact areas of order 10^{-14} m^2 and below, thus reducing band-edge shifting effects and making high resolution spatial profiling possible.

The greatest difficulty encountered with contact mode C-AFM has been tip wear and reliability. For UHV work, the normal forces necessary to get a stable electrical contact can be of the order 1 nN, whereas for work in air, it is typically hundreds of nN due to the presence of contaminants and oxide layers on the tip and sample. Consequently, metal-coated tips wear rather quickly in air, and measurements tend to be irreproducible. Investigations into the reliability of various tips have been carried out [74,75] clearly showing the effects of wear on tips due to lateral forces. Even single-crystal Si tips experience degradation under relatively low loads in UHV.

Initial experiments with C-AFM were made using metal-coated Si and Si_3N_4 probes, and etched W wire tips. Resistance maps of a metal sample were published apparently showing variations in resistance across a tungsten sample surface [76], as were Fowler–Nordheim maps of thin oxide layers [77]. If we look at the Fowler–Nordheim term in more detail, we find the current to be of the form

$$I = A_{\text{eff}} \frac{e^3}{8\pi h\phi} \left(\frac{V}{s} \right)^2 \exp\left(\frac{-8\pi\sqrt{2m_{\text{eff}}}}{3he} \frac{\phi^{3/2}s}{V} \right), \qquad (4.22)$$

where A_{eff} is the effective field-emission area, s is the oxide thickness, ϕ the electronic barrier height between the tip and the oxide, V is the potential difference between the tip and sample, and m_{eff} is the electronic effective mass within the oxide. This formula must be slightly modified for use in describing SPM experiments, as there will be a local enhancement of the electric field due to the sharpness of the tip. Clearly, as in all tunnelling processes, the current is strongly influenced by the oxide thickness. This technique has therefore been useful in mapping out variations in oxide thickness across ICs [78] and for studying oxide breakdown [79,80]. By fitting current–voltage curves to the above formula, the local work-function and oxide thickness may be inferred.

In certain situations, when the resistance of the sample is too high (≥ 10 GΩ) it is impractical to measure current flow, as there is too high a risk of causing oxide breakdown. An alternative is simply to measure the tip–sample capacitance. This has been used as a means to image trapped charge in oxide layers [81].

For practical applications it is important to be able to perform measurements under ambient conditions. As tip wear is a problem, and significantly higher forces are needed as compared to UHV conditions to penetrate contamination layers, it is desirable to have a tip made out of a single, conducting material. For UHV work, doped silicon tips may be used but an oxide forms under ambient conditions.

In order to overcome the bottleneck of tip wear, it is possible to use either harder tips or a technique which will reduce the lateral forces encountered during scanning in contact mode. In fact, both solutions now exist.

For contact mode measurements, extremely hard, high-conductivity tips have been developed, in the form of highly doped diamond coatings [82].

With the advent of the diamond-coated tip, it has been possible to perform high-resolution, reproducible potential measurements under ambient conditions. To illustrate this, we will now consider a number of applications of this technique as applied to actual devices. The first example is the use of C-AFM to gain insight into the failure mechanisms of current-stressed nanowires, a topic which we will consider in detail in Chapter 5. Electron microscopy usually cannot tell us about the microstructure at this scale, so we can use AFM combined with a conducting tip. In this way, we can simultaneously image the topography and the potential distribution along a current-carrying wire [83]. The high resolution of AFM enables us to image the microstructure. Figure 4.18 shows a potential map of a wire showing the linear drop in potential from one end to the other.

As we increase the current we can observe changes in the form of the potential drop as it changes from linear to step-like. This is due to

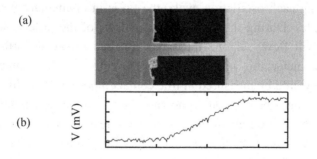

Fig. 4.18. (a) C-AFM map of potential along a 1 μm long gold nanowire, across which 100 mV was applied, showing (b) a linear drop in potential along the wire, along the line shown in (a).

electromigration and Joule heating induced restructuring of grain boundaries, causing the resistance to increase.   At points of high resistance, the local Joule heating increases and the atomic diffusivity increases, both of which combine to speed up the rate of electro-migration and hence failure.   For this reason, failure typically occurs along grain boundaries.   An example of this is shown in Fig. 4.19, where we see the topography and potential map of a nanowire which has just failed along a grain boundary due to electromigration.

(a)   (b)

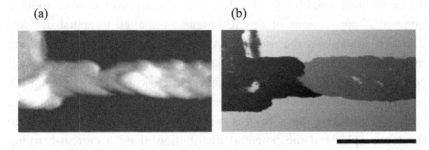

Fig. 4.19.   (a) Topography and (b) C-AFM map of surface potential along a gold nanowire which has failed due to electromigration.   Scale bar = 200 nm.

We mentioned earlier that there is a conducting AFM technique whereby one can reduce lateral forces during scanning and hence increase tip lifetime.   This is based on tapping mode, and uses that fact that during each oscillation cycle, the tip is in momentary contact with the sample.   During this time (which will be of the order a few μs), a current will flow.   If one measures the time-averaged current with a simple ammeter, this will register as a DC current, albeit a much smaller one than would be measured in true contact mode.   Again, this current is limited using a resistor, and one can extract the surface potential from the current measurement by choosing a resistor which is orders of magnitude larger than the *time averaged* tip–sample resistance.   As an example, if we use a tip which has a DC contact resistance to a sample of 1 kΩ, and we assume that in tapping mode a current flows for 1% of the cycle time, then the average tip–sample resistance is around 100 kΩ.

A limiting resistor of 10 MΩ will then allow us to measure the surface potential with an error of approximately 1%. The use of a larger resistance will only serve to improve this, as long as a sensitive enough current amplifier is used. An example of a potential map obtained in this manner is shown in Fig. 4.20 [84].

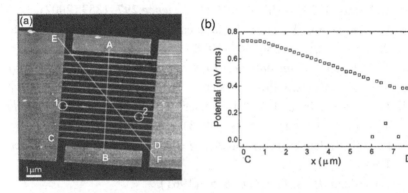

Fig. 4.20. (a) Topography and (b) C-AFM map of surface potential along a gold nanowire [84].

In summary, scanning probe techniques may be used to investigate and measure the electrical properties of nanostructures, devices and molecules. The family of scanning probe multimeters is an ever-growing one of increasing importance as device scales decrease and novel nanostructures are being fabricated.

References for Chapter 4

1. S. Datta, *Electronic Transport in Mesoscopic Systems* (Cambridge University Press, Cambridge, 1995).
2. D. K. Ferry and S. M. Goodnick, *Transport in Nanosctructures* (Cambridge Univesity Press, Cambridge, 1997).
3. M. J. Kelly, *Low Dimensional Semiconductors* (Oxford University Press, New York, 1995).
4. G. Binnig and H. Rohrer, *Helv. Phys. Acta* **55**, 726 (1982).

5. M. F. Crommie, C. P. Lutz and D. M. Eigler, *Science* **262**, 218 (1993).
6. M. F. Crommie, C. P. Lutz, D. M. Eigler and E. J. Heller, *Surf. Rev. Lett.* **2**(1), 127–137 (1995).
7. A. J. Heinrich, C. P. Lutz, J. A. Gupta and D. M. Eigler, *Science* **298**, 1381 (2002).
8. N. Nilius, T. M. Wallis and W. Ho, *Science* **297**, 1853 (2002).
9. V. Madhavan, W. Chen, T. Jamneala, M. F. Crommie and N. S. Wingreen, *Science* **280**, 567 (1998).
10. C. J. Chen, *An Introduction to Scanning Tunnelling Microscopy* (Oxford University Press, 1993).
11. J. A. Stroscio and W. J. Kaiser (eds.), *Scanning Tunnelling Microscopy* (Academic Press, 1993)
12. R. Wiesendanger (ed.), *Scanning Probe Microscopy* (Springer, 1998).
13. J. Bardeen, *Phys. Rev. Lett.* **6**, 57 (1961).
14. J. Tersoff and D. R. Hamann, *Phys. Rev. Lett.* **50**, 1988 (1993).
15. Y. Kuk, P. J. Silverman and F. M. Chua, *J. Microscopy* **152**, 449 (1988).
16. F. Jensen, F. Besenbacher, E. Laegsgaard and I. Stensgaard, *Phys. Rev. B* **41**, 10233 (1991).
17. R. S. Becker, J. A. Golovchenko, E. G. McRae and B. S. Swartzentruber, *Phys. Rev. Lett.* **55**, 2028 (1985).
18. R. J. Hamers and J. E. Demuth, *Phys. Rev. Lett.* **56**, 1972 (1986).
19. R. S. Becker, J. A. Golovchenko, D. R. Hamann and B. S. Swartzentruber, *Phys. Rev. Lett.* **55**, 2032 (1985).
20. R. M. Feenstra, A. J. Slavin, G. A. Huld and M. A. Lutz, *Phys. Rev. Lett.* **66**, 3257 (1991).
21. J. A. Stroscio, R. M. Feenstra and A. P. Fein, *Phys. Rev. Lett.* **57**, 2579 (1986).
22. K. Takayanagi and Y. Tanishiro, *Phys. Rev. B* **34**, 1034 (1986).
23. K. McElroy, J. Lee, J. A. Slezak, D.-H. Lee, H. Eisaki, S. Uchida and J. C. Davis, *Science* **309**, 1048 (2005).

24. K. McElroy, D. H. Lee, J. E. Hoffman, K. M. Lang, J. Lee, E. W. Hudson, H. Eisaki, S. Uchida and J. C. Davis, *Phys. Rev. Lett.* **94**, 197005 (2005).

25. T. Hanaguri, C. Lupien, Y. Kohsaka, D.-H. Lee, M. Azuma, M. Takano, H. Takagi and J. C. Davis, **430**, 1001 (2004).

26. S. H. Pan, E. W. Hudson and J. C. Davis, *Appl. Phys. Lett.* **73**, 2992 (1998).

27. S.-W. Hla, L. Bartels, G. Meyer and K.-H. Rieder, *Phys. Rev. Lett.* **85**, 2777 (2000).

28. C. Joachim, H. Tang, F. Moresco, G. Rapenne and G. Meyer, *Nanotechnology* **13**, 330 (2002).

29. A. Wachowiak, R. Yamachika, K. H. Khoo, Y. Wang, M. Grobis, D.-H. Lee, S. G. Louie and M. F. Crommie, *Science* **310**, 5747 (2005).

30. W. Ho and F. Hahn, *Phys. Rev. Lett.* **87**, 166102 (2001).

31. N. P. Guisinger, M. E. Greene, R. Basu, A. S. Baluch and M. C. Hersam, *Nano Lett.* **4**, 55 (2004).

32. A. Wachowiak, J. Wiebe, M. Bode, O. Pietzsch, M. Morgernstern and R. Wiesendanger, *Science* **298**, 577 (2002).

33. O. Pietzsch, A. Kubetzka, M. Bode and R. Wiesendanger, *Science* **292**, 2053 (2001).

34. S. Heinze, M. Bode, A. Kubetzka, O. Pietzsch, X. Nie, S. Blügel and R. Wiesendanger, *Science* **288**, 1805 (2000).

35. M. Bode, S. Heinze, A. Kubetzka, O. Pietzsch, X. Nie, G. Bilhmayer, S. Blügel and R. Wiesendanger, *Phys. Rev. Lett.* **89**, 237205 (2002).

36. A. Kubetzka, M. Bode, O. Pietzsch and R. Wiesendanger, *Phys. Rev. Lett.* **88**, 057201 (2002).

37. N. A. Pradhan, N. Liu and W. Ho, *J. Phys. Chem.* **109**, 8513 (2005).

38. N. Liu, N. A. Pradhan and W. Ho, *J. Chem. Phys.* **120**, 11371 (2004).

39. X. H. Qiu, G. V. Nazin and W. Ho, *Phys. Rev. Lett.* **92**, 206102 (2004).

40. B. C. Stipe, M. A. Rezaei and W. Ho, *Science* **280**, 1732 (1998).
41. P. Martensson and R. M. Feenstra, *Phys. Rev. B* **39**, 7744 (1989).
42. R. Wiesendanger, M. Bode, R. Pascal, W. Allers and U. D. Schwarz, *J. Vac. Sci. Technol. A* **14**, 1161 (1996).
43. A. Rettenberger, C. Baur, K. Läuger, D. Hoffmann, J. Y. Grand and R. Möller, *Appl. Phys. Lett.* **67**, 1217 (1995).
44. L. Olesen, M. Brandbyge, M. R. Sorensen, K. W. Jacobsen, E. Laegsgaard, I. Stensgaard and F. Besenbacher, *Phys. Rev. Lett.* **76**, 1485 (1996).
45. Z. Y. Rong and P. Kuiper, *Phys. Rev. B* **48**, 17427 (1993).
46. N. D. Lang, *Phys. Rev. B* **37**, 10395 (1988).
47. E. Tekman and S. Ciraci, *Phys. Rev. B* **43**, 7145 (1991).
48. P. Muralt and D. W. Pohl, *Appl. Phys. Lett.* **48**, 514 (1986).
49. M. A. Schneider, M. Wenderoth, A. J. Heinrich, M. A. Rosentretter and R. G. Ulbrich, *Appl. Phys. Lett.* **69**, 1327 (1996).
50. B. G. Briner, R. M. Feenstra, T. P. Chin and J. M. Woodall, *Phys. Rev. B* **54**, 5283 (1996).
51. F. M. Battiston, M. Bammerlin, C. Loppacher, R. Lüthi, E. Meyer, H.-J. Güntherodt and F. Eggimann, *Appl. Phys. Lett.* **72**, 25 (1998).
52. D. Hoffmann, A. Haas, T. Kunstmann, J. Seifritz and R. Möller, *J. Vac. Sci. Technol. A* **15**, 1418 (1997).
53. C. C. Williams and H. K. Wickramasinghe, *Nature* **344**, 317 (1990).
54. G. Binnig, C. F. Quate and Ch. Gerber, *Phys. Rev. Lett.* **56**, 930 (1986).
55. T. R. Albrecht and C. F. Quate, *J. Vac. Sci. Technol. A* **6**, 271 (1988).
56. G. Meyer and N. M. Amer, *Appl. Phys. Lett.* **53**, 1044 (1988).
57. W. Steinhogel, G. Schindler, G. Steinlesberger and M. Engelhardt, *J. Appl. Phys.* **97**, 23706 (2005).
58. U. D. Schwarz, W. Allers, G. Gensterblum and R. Wiesendanger, *Phys. Rev. B* **52**, 14976 (1995).
59. H. K. Wickramasinghe, *J. Vac. Sci. Technol. A* **8**, 363 (1990).

60. M. A. Lantz, H. J. Hug, P. J. A. van Schendel, R. Hoffmann, S. Martin, A. Baratoff, A. Abdurixit, H.-J. Guntherodt and Ch. Gerber, *Phys. Rev. Lett.* **84**, 2642 (2000).

61. C. Durkan, M. E. Welland, D. P. Chu and P. Migliorato, *Phys. Rev. B* **60**, 16198 (1999).

62. H. O. Jacobs, P. Leuchtmann, O. J. Homan and A. J. Stemmer, *J. Appl. Phys.* **84**, 1168 (1998).

63. J. M. R. Weaver and D. W. Abraham, *J. Vac. Sci. Technol. B* **9**, 1559 (1991).

64. R. Shikler, T. Meoded, N. Fried and Y. Rosenwaks, *Appl. Phys. Lett.* **74**, 2972 (1999).

65. J. R. Barnes, A. C. F. Hoole, M. P. Murrell, M. E. Welland, A. N. Broers, J. P. Bourgoin, H. Biebuyck, M. B. Johnson and B. Michel, *Appl. Phys. Lett.* **67**, 1538 (1995).

66. H. O. Jacobs, PhD thesis, Swiss Federal Institute of Technology, Zurich (1999).

67. A. S. Hou, F. Ho and D. M. Bloom, *Forces in Scanning Probe Methods*, eds. H.-J. Güntherodt, D. Anselmetti and E. Meyer (Kluwer, Dordrech, 1995), p. 640.

68. C. Böhm, J. Sprengepiel, M. Otterbeck and E. Kubalek, *J. Vac. Sci. Technol. B* **14**, 842 (1996).

69. M. A. Lantz, S. J. O'Shea and M. E. Welland, *Phys. Rev. B* **56**, 15345 (1997).

70. R. Fowler and L. Nordheim, *Proc. R. Soc. Lond. A* **119**, 173 (1928).

71. K. L. Johnson, *Contact Mechanics* (Cambridge University Press, Cambridge, 1985).

72. P. De Wolf, T. Clarysse, W. Vandervorst and L. J. Hellemans, *J. Vac. Sci. Technol. B* **16**, 401 (1998).

73. P. De Wolf, J. Snauwert, L. Hellemans, T. Clarysse, W. Vandervorst, M. D'Olieslaeger and D. Quaeyhaegens, *J. Vac. Sci. Technol. A* **13**, 1699 (1995).

74. F. Houzé, R. Meyer, O. Schneegans and L. Boyer, *Appl. Phys. Lett.* **69**, 1975 (1996).

75. M. A. Lantz, S. J. O'Shea and M. E. Welland, *Rev. Sci. Instrum.* **69**, 1757 (1998).

76. T. G. Ruskell, R. K. Workman, D. Chen, D. Sarid, S. Dahl and S. Gilbert, *Appl., Phys. Lett.* **68**, 93 (1996).

77. W. Steinhogel, G. Schindler, G. Steinlesberger and M. Engelhardt, *J. Appl. Phys.* **97**, 23706 (2005).

78. A. Olbrich, B. Ebersberger and C. Boit, *Appl. Phys. Lett.* **73**, 3114 (1998).

79. S. J. O'Shea, R. M. Atta, M. P. Murrell and M. E. Welland, *J. Vac. Sci. Technol. B* **13**, 1945 (1995).

80. M. E. Welland and M. P. Murrell, *Scanning* **15**, 251 (1993).

81. R. C. Barratt and C. F. Quate, *Ultramicroscopy* **42–44**, 262 (1992).

82. Nanosensors GmbH, Wacholderweg 8, D-71134, Aldlingen, Germany.

83. M. C. Hersam, A. C. F. Hoole, S. J. O'Shea and M. E. Welland, *Appl. Phys. Lett.* **72**, 915 (1998).

84. A. Bietsch, M. A. Schneider, M. E. Welland and B. Michel, *J. Vac. Sci. Technol. B* **18**, 1160 (2000).

Chapter 5

Electromigration: *How Currents Move Atoms, and Implications for Nanoelectronics*

5.1. Introduction to Electromigration, Wire Morphology

The scale of interconnects is already below 1 μm, and is rapidly approaching that at which surface effects will become prevalent. It is therefore becoming increasingly important to characterise and understand the electrical transport properties and failure mechanisms of ten-nanometer-scale metallic wires. Wires with such dimensions are mesoscopic, but when an electric current passes through them, they can thin down to become more like atomic point contacts. In this chapter, we wish to gain an understanding of the mechanisms by which material is transported by the flow of electric current ("electromigration"). The parameters which control this process are the diffusivity of the material from which the wire is made, the wire geometry, the current, the temperature, and the microstructure of the wire, which ultimately determines the net rate of diffusion of material along the wire.

Interconnect microstructure generally falls into three classes: (i) single-crystal, (ii) bamboo and (iii) polycrystalline. These are illustrated below in Fig. 5.1.

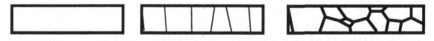

Fig. 5.1. Interconnect microstructure, (i) single-crystal, (ii) bamboo, (iii) polycrystalline.

5.2. Fundamentals of Electromigration — The Electron Wind

Conventional current carrying wires eventually fail due to a process known as electromigration [1–5], which is the thermally assisted motion of ions under the influence of an electric field. When a wire is supporting a current, it heats up due to the electron–phonon interaction, and the metal atoms become more mobile. The atoms move against the direction of current flow, propelled by the "electron wind", with a rate determined by the local diffusion coefficient, as illustrated in Fig. 5.2. This is an activated diffusion process, so the rate increases with increasing temperature. Generally the devices to which interconnects are connected carry smaller current densities than the wires themselves giving rise to lower Joule heating there, and therefore they act as heat sinks. As a consequence, material flowing along such a wire is not replaced at the ends at the same rate as it leaves, and voids form at the negative end (cathode). These voids may eventually merge, resulting in failure. As mass transport progresses, mechanical stresses (compressive) develop within the wire, setting up a stress gradient which opposes electromigration. Eventually the stresses reach a critical value at which point the wire fails.

It has been found that the mean amount of time it takes a wire to fail for a given applied current density is governed by the following equation, known as Black's equation [6]:

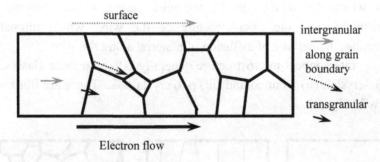

Fig. 5.2. Motion of ions due to the flow of electric current. The length of the respective arrows indicates the relative diffusion rates in each case.

$$t_{50} = CJ^{-n}e^{E_a/k_BT},$$ (5.1)

where t_{50} is the mean time to failure, C is a constant factor dependent on the material properties (geometry, diffusion constant, etc.), J is the current density, T is the temperature, E_a is the electromigration activation energy (0.5–0.7 eV for Al [7]) and n is an integer which in most cases is 2.

Clearly then, the lifetime of wires scales inversely with the current density squared, and inverse exponentially with the temperature. As was shown by Blech [8], the velocity with which material is transported along a wire by electromigration depends inversely on the wire length. There is also a critical length (*Blech length*), dependent on the current density, below which electromigration is entirely suppressed. This is considered to be due to the balance between the electron wind force and the backflow of material induced by the compressive stress gradient which arises when material builds up at the cathode. For a fixed line length then, as the current density is increased above the threshold level, the compressive stress increases to the point where plastic flow occurs. This yield stress depends on wire size, and increases as wires get smaller ($< 1\ \mu$m), mostly due to the presence of fewer defects.

Electromigration is generally described phenomenologically in terms of the evolution of stress, given by:

$$\frac{\partial \sigma}{\partial t} = \frac{\partial}{\partial x}\left[\kappa\left(\frac{\partial \sigma}{\partial x} + G\right)\right],$$ (5.2)

where $\kappa = DB\Omega/k_BT$ and G is the electromigration driving force, $G = Eq'/\Omega$. Here, D is the effective atomic diffusivity, B is the Young's modulus of the material consisting the wire, Ω is the atomic volume, E is the electric field, q' is the effective atomic charge and k_B, T are Boltzmann's constant and the temperature, respectively. In studies of electromigration, the aim is to find solutions to the above equation under the appropriate boundary conditions. A number of authors have performed detailed calculations of the evolution of stress under various conditions in order to predict the geometric and material dependence of electromigration failure [9–13] with relatively good agreement between theory and experiment. Here we will consider analytical solutions to the stress equation for a nanowire device.

5.3. Electromigration-Induced Stress in a Nanowire Device

Solving the above equation, we obtain for the distribution of stress in a finite line of length $2L$ with the centre at $x = 0$:

$$\sigma = 2GL\left(-\frac{x}{2L} - 4\sum_{n=0}^{\infty}\left(\frac{1}{\eta^2}e^{-\eta^2 t/\tau}\cos\left(\eta\frac{x+L}{2L}\right)\right)\right), \qquad (5.3)$$

where $\eta = (2n+1)\pi$ and $\tau = 4L^2/\kappa$, which is the characteristic time taken for the stress to reach a steady level. Due to the relationship between the electric field and the current density, i.e. $E = J\rho$ where ρ is the electrical resistivity, we can see that there is a linear relationship between the stress in the wire and the current density. Up to a time of the order τ, the stress develops exponentially, and thereafter it increases linearly with time. For a fixed temperature, this time should essentially be a constant irrespective of J. Failure is considered to occur once the stress reaches some critical value. For typical values quoted for Al of $D = 3 \times 10^{-16}$ m^2/s for bamboo structures and 6×10^{-14} m^2/s for polycrystalline structures, $B = 78$ GPa, $L = 0.5$ μm, $\Omega = 1.7*10^{-29}$ m^3 and $q' = -6.4 \times 10^{-19}$ C we obtain a time constant $\tau \sim 2$ ms for bamboo wires (i.e. where the wire has only one grain across its width) and 0.5 ms for polycrystalline wires (i.e. many grains across the width), this time decreasing with increased temperature. Representative graphs of stress versus position and time are shown in Figs. 5.3 and 5.4, respectively. Consequently, when measuring the mean time to failure (MTF), one should maintain the current density and temperature low enough to ensure that failure occurs in the region where stress is increasing linearly with time. Otherwise, due to the large rate of increase in stress for times below this, MTF measurements for higher current densities would have

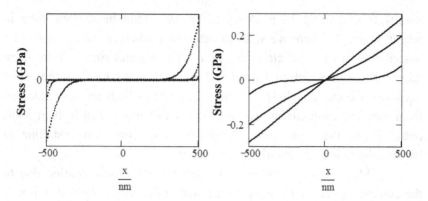

Fig. 5.3. Stress versus position for bamboo (dotted curves) and polycrystalline (solid curves) wires of length 1 μm, width 50 nm, height 20 nm, for a current of 1.8 mA, for times of 1, 10 and 100 ms.

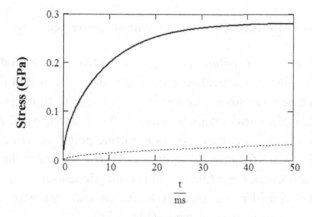

Fig. 5.4. Peak stress (in GPa) versus time for bamboo (dotted curve) and polycrystalline (solid curve) wires of length 1 μm, width 50 nm, height 20 nm, for a current of 1.8 mA.

large errors associated with them. For this reason, MTF measurements are typically performed over a timescale of hours or even days.

To date, the smallest interconnect structures studied both experimentally and theoretically have been of the order 1 μm and larger in width, for which case there has been rather good agreement between experiment and theory. Experimental studies [14,15] on wires with widths from around 1 μm down to below 20 nm have been done in recent years, and here we wish to determine whether failure should be describable in terms of electromigration and thermal stress, or are there other influences which play a role? From the simple result in the above equation, we can conclude a number of points which are also obtained from detailed calculations: (i) long wires fail more readily than short ones, (ii) as the wire width shrinks to pass from polycrystalline to bamboo-like, the wire should fail less easily.

What we have not yet considered is that Joule heating due to the current will cause the temperature within the wire to be non-uniform, modifying the distribution of stress. In order to incorporate this to a first approximation into the above equation for the stress, we will now look at calculations of the temperature distribution within a wire.

5.4. Current-Induced Heating in a Nanowire Device

The rate of failure due to electromigration depends on the temperature (it is an activation process), which in turn will depend on the balance between the net energy input to the wire from current flow, i.e. $J^2\rho$, (J is the current density, and ρ is the resistivity) and the rate of energy loss to the surroundings. As a starting point, we will construct a simple 1D model for heat dissipation in a wire and infer the expected failure characteristics as a function of the wire dimensions.

We consider the case of a current-carrying wire connected between two semi-infinite heatsinks (Fig. 5.5).

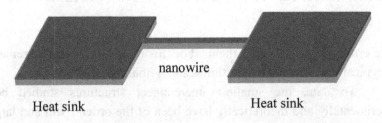

Fig. 5.5. Schematic of device structure used in resistive heating calculations.

The equation describing the steady-state excess temperature (T) is Poisson's equation, where the source term is given by the position-independent rate of generation of heat energy, Q/k and the losses due to heat flow through the substrate i.e. [16]:

$$\nabla^2 T - m^2 T + \frac{Q}{k} = 0, \qquad (5.4)$$

where

$$m = \sqrt{\frac{k_{sub}}{ktd}},$$

and $Q = J^2\rho$; and k, k_{sub}, t and d are the thermal conductivity and thickness of the wire and the substrate, respectively. It should be noted that the resistivity of such a film will be approximately 2.5 times larger than the bulk value due to grain boundary scattering and the Fuchs size effect, as we saw in Chapter 3. The thermal conductivity is likewise reduced in accordance with the Wiedemann–Franz law [17]. The measured resistivity of 20 nm thick Au films is typically ~6 $\mu\Omega$-cm, as compared to the bulk value of 2.4 $\mu\Omega$-cm at room temperature. The wire length is L, and the centre of the wire is at $x = 0$. The bulk thermal conductivities of gold and SiO$_2$ are 300 and 1.1 W.K^{-1}m^{-1} [18]. We must assume that the thermal conductivity of the wires is scaled down from the bulk value by a factor of 6/2.4, and that the actual value we need to use for SiO$_2$ is 0.55 W.K^{-1}m^{-1}, as the wire is only surrounded on one side by SiO$_2$.

The assumptions we make are (i) the heat flow is one-dimensional, along the "x" axis, (ii) the current density in the contacts is several orders of magnitude smaller than in the wire, so we can neglect the heat generation term within the contacts, (iii) consequently, we assume the current density is described by a step function, (iv) there are no "hot spots" within the wire, i.e. localised points of higher resistance than the surroundings, (v) the only heat losses are through the substrate (convective and radiative losses to the surroundings should be orders of magnitude less efficient), (vi) the resistance of the wire does not change with temperature. Assumptions (iii) and (vi) will provide the greatest departure from reality as the current density will have a smooth transition from the wire to the contact, the slope decreasing with increasing wire width, and the resistance will change by around 0.1% per degree. Later in this chapter, we will analyse the effect of a temperature-dependent resistance.

The boundary conditions which determine the solution of Poisson's equation are that at the wire-contact junction, the temperature profile is continuous, as is its first derivative, and at the centre of the wire, the *net* directional flow of heat is through the substrate, as equal amounts of heat will flow towards both ends away from the centre, due to the symmetry. The rate of heat loss is proportional to the wire width as that defines the contact area with the substrate through which heat leaks out, but as the rate of heat generation has the same dependency, the net width dependence cancels out. We will see later that there is in fact a net width dependence due to the fact that the resistivity of a nanowire will depend on size once its dimensions are comparable to the mfp of the conduction electrons.

We find the solution of Poisson's equation in the wire and the contacts to be:

$$T_{\text{wire}} = -\frac{Q}{2km^2}e^{-mL}\left(e^{mx}+e^{-mx}\right)+\frac{Q}{km^2},$$

$$T_{\text{contact}} = Q\left[\frac{1}{2km^2}\left(e^{mL}-e^{-mL}\right)\right]e^{-m|x|}.$$

(5.5)

The salient points are (1) the temperature is a maximum at the centre of the wire, (2) the temperature gradient is a maximum at the wire-contact junction, (3) for a given current density, the temperature is dependent on the wire thickness and length, and not on the width. As argued above the narrower the wire is as compared to the contacts, the better assumption (iii) is satisfied. In Fig. 5.6, the calculated temperature of a 50 nm wide, 20 nm thick and 1000 nm long wire carrying currents of 1, 1.5 and 2 mA is shown. We can see from Fig. 5.6 that a relatively modest current will cause such a wire to reach high temperatures, which will accelerate the electromigration process. The temperature depends very strongly on the efficiency of heat dissipation through the substrate, as we are neglecting all other forms of heat loss.

Figure 5.7 shows the peak temperature at the centre of a wire, as a function of the underlying oxide thickness. As one might expect, the temperature is extremely sensitive to the oxide thickness, wherein lies

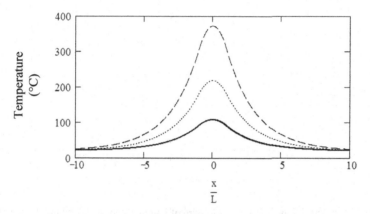

Fig. 5.6. Temperature profile for a gold wire of length 1 μm, width 50 nm, height 20 nm, on a 100 nm thick layer of SiO$_2$, for a current of 1 mA (solid curve), 1.5 mA (dotted curve) and 2 mA (dashed curve).

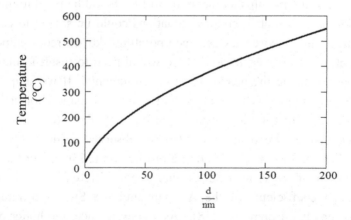

Fig. 5.7. Peak temperature for a gold wire of length 1 μm, width 50 nm, height 20 nm, current of 2 mA, as a function of underlying SiO$_2$ thickness.

the somewhat obvious conclusion — to help wires last longer, maximise the heat sinking through the substrate.

In Fig. 5.8, we have plotted the temperature distribution in a wire for several different oxide thicknesses, from which we can see that as the thickness decreases and more heat is lost to the substrate, the temperature becomes more uniformly distributed along the wire.

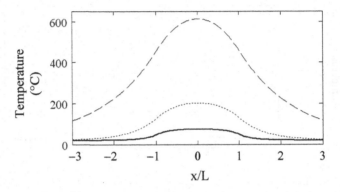

Fig. 5.8. Temperature profile for a gold wire of length 1 μm, width 50 nm, height 20 nm, for a current of 2 mA, on a layer of SiO_2 of thickness 100 nm (solid curve), 20 nm (dotted curve) and 2 nm (dashed curve).

In principle, this temperature profile should be input to the stress equation, and the set of coupled equations could be solved to calculate the stress in a wire device, incorporating the correct temperature distribution. However, to do that we would need to resort to numerical methods, reducing the generality of our arguments. If we simply input the above expressions for temperature into our equation for stress, we obtain a stress distribution as shown in Fig. 5.9. The stress is increased throughout the wire relative to what we observed earlier with a static stress distribution. The temperature profile can tell us about the spatial distribution of stress in a device due to the difference in the thermal expansion coefficients of the Au wire and the SiO_2 substrate, α of approximately 14 ppm.$^\circ$C^{-1} [18]. As a consequence, the hotter parts of the wire experience greater thermomechanical compressive stresses. As our model tells us that the temperature is greatest at the centre, this means the thermal stress $(-3\alpha B\Delta T)$ is greatest at the centre, as opposed to the electromigration stress which is greatest at the ends.

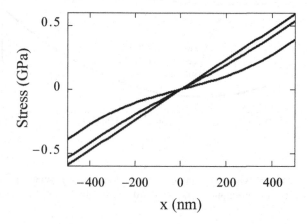

Fig. 5.9. Stress versus position for a polycrystalline Al wire of length 1 μm, width 50 nm, height 20 nm, for a current of 1.8 mA, at 1, 2.5 and 10 ms. This is taking into account that temperature is a function of position, to first order.

The total stress, which is the sum of these two terms, is plotted in Fig. 5.10 for a number of currents. What we can see is that for low currents, the stress distribution is linear along the wire, reaching a maximum at the wire ends, but as the current is increased, the thermomechanical stress starts to become appreciable, and causes the peak to shift towards the centre of the wire.

Of course, this has been all on the assumption that we have so many grains across the wire width and thickness that the effective diffusion constant for the wire is a global constant, rather than a local variable.

In reality, interconnects typically have a bamboo microstructure, and the stress distribution will be discontinuous, being approximately linear within each grain, and peaking at each grain boundary, as shown in Fig. 5.11. The reason for the stress peaking at each boundary is because the diffusion constant changes rapidly there.

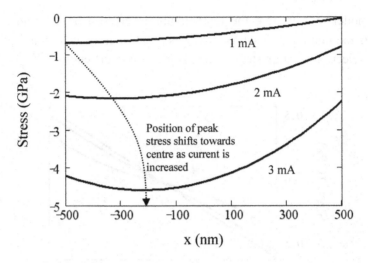

Fig. 5.10.   Net stress (electromigration + thermomechanical) versus position for a polycrystalline Au wire of length 1 μm, width 50 nm, height 20 nm, for currents of 1, 2 and 3 mA.

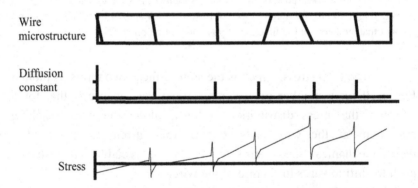

Fig. 5.11. Stress in a bamboo wire.

What we can conclude from this semi-quantitative approach is that an ideal interconnect (from the standpoint of stability against electromigration) should be bamboo-type, with a large number of grain boundaries spanning the width of the wire, as that will have the lowest net rate of diffusion.

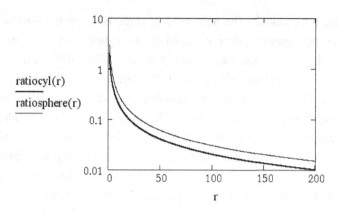

Fig. 5.12. Ratio of surface area to volume for a cylinder (thick line) and a sphere (thin line) as a function of radius.

5.5. Diffusion of Material, Importance of Surfaces, Failure of Wires

One of the key reasons behind the development of nanotechnology lies in the fact that nanostructures possess fundamentally different properties to larger-scale structures. In many instances, this is either due to the fact that nanostructures by their very nature have a large surface area to volume ratio, as illustrated in Fig. 5.12 for spherical and cylindrical structures, or that they are small enough to exhibit quantum size effects. Surfaces are different from the bulk in a number of important ways, in particular their electronic states and phenomena associated with them, e.g. electrical properties, optical response, plasmon resonances; and also the diffusivity of surface atoms tends to be much greater than that of atoms within the bulk.

From a qualitative viewpoint then, we should consider what will happen to an interconnect as we shrink its width, whilst maintaining a constant mean grain size. First of all, when the wire width shrinks below the mean grain diameter, the wire's microstructure changes from polycrystalline to bamboo. This will reduce the net rate of diffusion of material along the wire, so will be more stable. Then, as the wire width decreases below the electronic mfp, the resistivity will start to increase above the bulk value. The next critical stage is when the wire's surface area becomes appreciable compared to its volume, as then the rate of failure will start to increase again due to surface diffusion. We can describe this mathematically in the following way — consider a wire of width w, thickness t, and mean grain diameter δ. The two fastest routes for diffusion of atoms are on the surface and along grain boundaries, so we can neglect the other diffusion pathways in this simple analysis. The drift velocity of atoms will be given by the diffusivity multiplied by a characteristic length. In the case of grain-boundary diffusion, the characteristic length is the average grain size (compared to the wire width), whereas for surface diffusion, the characteristic lengths are the wire width and thickness. We can write this in the following way, remembering that the wire is sitting on a surface, so atoms can only diffuse along the two sides and the top, and not on the bottom surface [19]:

$$\text{Grain boundary drift velocity} = \frac{1}{\delta}\left(1 - \frac{\delta}{w}\right)D_{\text{GB}} \text{ when } w > \delta, \text{ otherwise } 0.$$

$$\text{(5.6)}$$

$$\text{Surface drift velocity} = \left(\frac{1}{w} + \frac{2}{t}\right)D_S.$$

The total drift velocity of the diffusing atoms is therefore more or less dominated by the grain-boundary drift velocity when the wire width is larger than δ, and by the surface drift velocity otherwise. This is illustrated in Fig. 5.13, where we plot both the dependence of these drift velocities as well as their sum (which will be proportional to the net rate of mass transport), on the wire width.

Of course, atoms will be diffusing through the wire as well as on the surfaces, and the quantity we are really interested in is the atom flux, which will be dependent on the drift velocity, and in this case, on the sum of both velocities above. We can see that the total drift velocity has a minimum when there is a cross-over from polycrystalline to bamboo.

The greater the net atom flux, the shorter will be the wire's lifetime, so we have plotted the inverse of the drift velocity in Fig. 5.14.

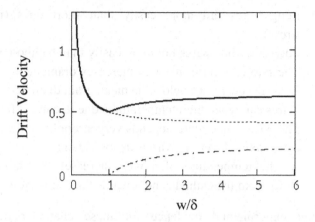

Fig. 5.13. Drift velocity along surface (dotted line), grain boundary (dashed line) and their sum (solid line) as a function of normalised wire width, w/δ.

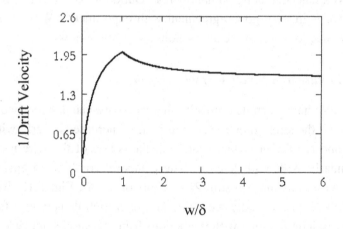

Fig. 5.14. 1/(drift velocity) as a function of normalised wire width, w/δ.

5.6. Experimental Observations of Electromigration and Heating in Nanowires

To summarise the points we have already deduced from our semi-quantitative look at electromigration and thermal stresses in nanowires, we have determined the following:

- Long wires fail more easily than short ones (increased stress);
- Polycrystalline wires fail more easily than bamboo wires;
- The rate of failure of wires increases dramatically once the wire size decreases below the mean grain diameter;
- Wires can reach high temperatures due to current flow;
- The wire temperature depends very strongly on the thickness of the oxide layer the wire is deposited on;
- As the temperature of a wire increases, the failure point shifts from the cathode end towards the centre of the wire.

What about experimental evidence for these characteristics? This has been addressed by many researchers for large scale ($> 1 \ \mu$m) interconnects (which we will look at briefly in Section 5.7), whereas very little work has been done on nanowires. Durkan has shown that much of the above is in very good agreement with experiment. We will look at the evidence for each of the above points separately now.

5.6.1. *Failure as a function of wire length*

We have seen theoretically that as a wire gets longer, two things happen: (i) the stress due to electromigration increases linearly with wire length and (ii) the peak temperature at the centre of the wire increases, also linearly with wire length. These factors combine to ensure that longer wires fail more readily than short ones. In Fig. 5.15, we have plotted the experimentally measured failure current density as a function of the wire length, along with predictions from our model above.

Clearly, there is excellent agreement, and the shorter wires can sustain larger current densities than longer wires. In Fig. 5.16, SEM images of the wires show that in general, short wires tend to fail along relatively short cracks.

5.6.2. *Failure as a function of wire width*

We have seen that of the two factors which contribute to wire failure: temperature and atom flux due to the electron wind, temperature

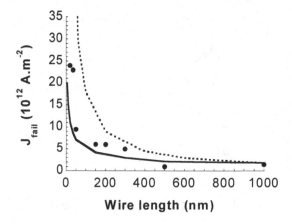

Fig. 5.15. Failure current density as a function of wire length, for gold wires 20 nm thick and 60 nm wide. Points are measured values, the curve is theoretical.

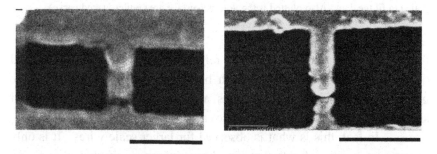

Fig. 5.16. SEM images of short nanowires after failure. Scale bars are 100 nm.

is independent of wire width, but the atom flux depends very strongly on wire width once the width is comparable to the mean grain size. In Fig. 5.17, we have plotted the failure current density as a function of wire width for wires between 20 nm and 850 nm long, along with predicted results on the basis of the diffusion model discussed earlier.

Again, there is excellent agreement between theory and experiment, the only fitting parameter used in the theory (apart from a scaling factor to account for the diffusivity of the nanowire material) is the mean grain size, which in this particular case was 100 nm. Whilst

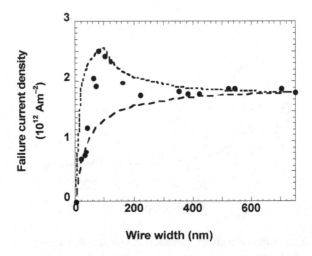

Fig. 5.17. Failure current density as a function of wire width, for gold wires 20 nm thick and 1 μm long. The points are measured values, the dotted curve is 1/(drift velocity), scaled to fit the data. Also plotted is the volume/surface area ratio (dashed curve).

this agreement is remarkable, if one remembers that the atom flux should decrease as soon as the wire width becomes smaller than the average grain size (i.e. when the wire has a bamboo-like microstructure), it should be the case that narrower wires are more stable than wider ones.

In fact, this is what is observed for larger scale wires. It is only because of surface effects that we observe the opposite behaviour. To further illustrate this point, we have plotted the volume/surface area ratio in Fig. 5.17, as this will show the same general dependence as the failure current density. An interesting point to note is that even though the wires are reaching a significant temperature and the thermomechanical stress is comparable to the electromigration stress, the failure characteristics can be largely explained in terms of electromigration.

In those cases where a wire is found to have failed catastrophically at the centre, we can conclude that it has failed by thermomechanical stress, and we can estimate the critical failure stress of the wire material using the peak temperature of the wire. We will see how to apply this with a specific example later.

5.7. Experimental Observations of Electromigration in Micron-Scale Wires

It has become standard practice to measure two quantities in particular when investigating electromigration in interconnects, (i) mean time to failure and (ii) critical length. Both measurements are typically performed using low enough current densities to minimise current-induced heating, and are performed at several different temperatures.

The mean time to failure measurements are typically performed as follows: wires are fabricated and their geometry precisely measured using SEM. They are then subjected to current densities large enough to cause the wire to fail over the timescale of a few days. This is repeated for several wires and several temperatures ranging from room temperature to around 250°C. Some examples are shown in Fig. 5.18.

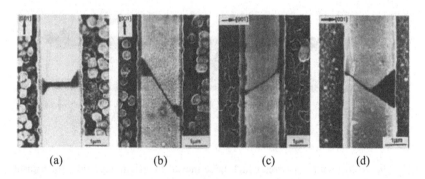

(a)　　　　　(b)　　　　　(c)　　　　　(d)

Fig. 5.18. SEM images of wires which have failed by electromigration. Wires are oriented along the (110) direction, with in-plane directions of [1$\overline{1}$0], [1$\overline{1}$2], [$\overline{1}$12] and [$\overline{1}$12]. (a)–(c) are unpassivated, whereas (d) is passivated.

Measurements of the critical length are typically performed using the so-called Blech block structure [8], whereby several rectangular metallic wires are deposited on top of a line of TiN, which is a refactory metal with a conductivity significantly lower than a metal. As a current is passed through the TiN line, as soon as it reaches one of the metal wires, the current by-passes the TiN and instead flows through the metal wire. In this way, a current can be made to flow through a wire without any contacts having to be made to it. The entire wire will move under the influence of the current, and by tracking this movement, information about the atomic drift velocity can be obtained, as well as the critical length below which no electromigration is observed. In Fig. 5.19, we show some typical results obtained in this manner.

(a)

(b)

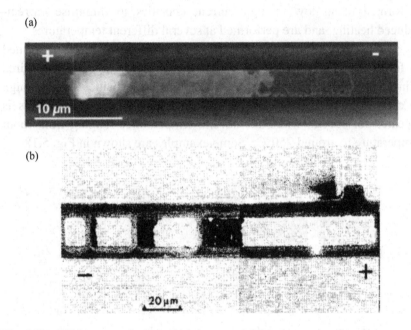

Fig. 5.19. SEM images of wires which have moved by electromigration, for measurements of the atomic drift velocity. This configuration is known as a Blech block. (a) and (b) are reproduced from Refs. 10 and 8, respectively.

5.8. Wire Heating — Additional Considerations

We have predicted that a wire may reach temperatures of several hundred degrees. The thermomechanical stress and the atom flux in electromigration depend on the temperature, which in turn depends on wire length, current density, oxide thickness and thermal conductivity of wire and oxide. We have already seen that the simple model presented above can explain the dependence of ease of failure on wire width and length. We already know that under certain conditions (i.e. when the current density is large enough) then the thermal stress is greater than the electromigration stress. A very simple way to test this is to recognise that because wires of any given geometry fail at a particular current density, and consequently a particular temperature, we can assign a critical failure temperature. This is just another way of saying that as we increase a wire's temperature, we will increase the thermal stress, and when this reaches a critical value, the wire will fail, so the concept of a critical failure temperature really means critical failure stress. There are two experiments that have been done to investigate this; the first of these is to alter the oxide thickness, and hence the temperature of a wire for a given current density, and look at the impact of this on the failure current density. This is shown in Fig. 5.20. Again, we can see excellent agreement between theory and experiment, demonstrating that the concept of critical failure temperature is indeed a useful one. One of the major assumptions that we have made so far is that the wire resistivity is independent of temperature. This will not be the case in reality, as we saw so clearly in Chapter 1. This will have the effect of increasing the resistance of a wire as the temperature is increased, so for a given current, the actual temperature of a wire will be higher than we have predicted, potentially by a significant amount. We saw in Chapter 1

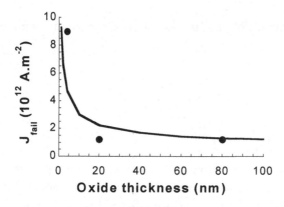

Fig. 5.20. Failure current density as a function of oxide thickness, for gold wires 20 nm thick, 1 μm long, 60 nm wide. The curve is based on a critical failure temperature of 220°C.

that the passage of a current can cause heating of a wire, which will cause the resistance to increase in the following manner:

$$R = \frac{R_0}{1 - \alpha\beta I^2},$$ (5.7)

where R_0 is the resistance at room temperature, and α and β are the temperature coefficient of resistance and the proportionality factor between power and temperature, respectively. If we take a typical nanowire device on a 20 nm thick oxide layer, the temperature at the centre is approximately 100 degrees. This gives a value for β of 4.055×10^5 degrees/W. For gold, $\alpha = 0.0083$ $\mu\Omega$-cm K^{-1}. If we therefore scale the resistivity ρ by this factor in Eq. (5.7) above, we find that the dependence of temperature on current deviates from what we saw earlier, and for a given injected current density, the temperature will indeed be higher than expected. This is shown in Fig. 5.21.

The difference in calculated temperatures increases with increasing temperature as one should expect, as then the resistance will be changing more rapidly, and for the case in point here, the difference is close to 40 degrees when the current is 2.5 mA. The only fitting parameter needed is the quantity β, which is the constant of proportionality between the input power to the nanowire and the temperature.

We can estimate β by determining the average temperature of a wire, i.e.

$$T_{\text{average}} = \frac{1}{2L} \int_{-L}^{L} T(x)dx.$$ (5.8)

β is then just $T_{\text{average}}/(I^2 R)$.

In Fig. 5.22, we have plotted the temperature profile along a wire using both a fixed resistance and a temperature-dependent resistance.

Knowing β, we can also predict the resistance as a function of current, as given by Eq. (5.7), which is shown in Fig. 5.23, and also the current as a function of voltage, as shown in Fig. 5.24.

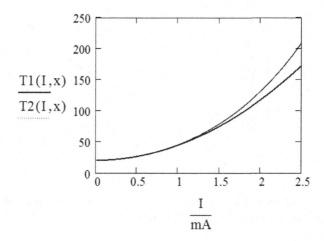

Fig. 5.21. Dependence of calculated peak temperature on current in a 60 nm wide, 20 nm thick, 1 μm long gold nanowire on a 20 nm thick SiO$_2$ layer (thick line), and taking into account the change of resistance with temperature (thin line).

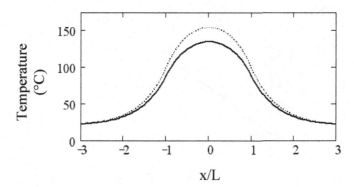

Fig. 5.22. Dependence of calculated temperature profile in a 60 nm wide, 20 nm thick, 1 μm long gold nanowire on a 20 nm thick SiO$_2$ layer (solid line), and taking into account the change of resistance with temperature (dotted line).

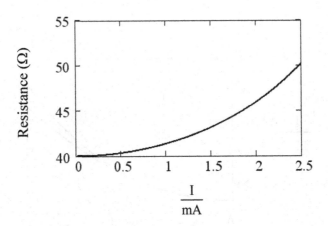

Fig. 5.23. Calculated dependence of resistance on current for a 60 nm wide, 20 nm thick, 1 μm long gold nanowire on a 20 nm thick SiO_2 layer.

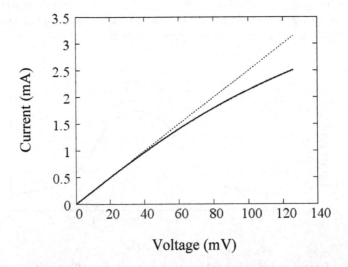

Fig. 5.24. Calculated dependence of current on voltage for a 60 nm wide, 20 nm thick, 1 μm long gold nanowire on a 20 nm thick SiO_2 layer. The dotted line is what one would obtain if the resistance were independent of temperature.

This demonstrates the fact that the resistance of a nanowire device can change by a large percentage for a relatively modest current, mainly due to the fact that a small current through a small structure can be equivalent to a very high current density. We are talking about currents of mA in a device with a cross-sectional area of around 10^{-15} m^2, which equates to 10^{12} A.m^{-2}. This is the same as passing a current of 10^6 A through a 1 mm wire, so the nanostructures we have been looking at so far can sustain truly enormous current densities, and it is no wonder that they can get so hot in the process. What about the experimental evidence for any of this? Well, we have seen that the concept of critical current density is tantamount to a critical failure temperature, which in turn, relates to the critical thermomechanical stress at which a wire will fail. How about more direct evidence for wire heating? We have just seen how wire heating will cause the resistance to increase, and we have calculated by how much it should change for a given wire sample. Figure 5.25 shows the experimentally measured change in resistance of a wire of similar dimensions, and the inferred temperature change assuming that $\alpha = 0.0083$ $\mu\Omega$-cm K^{-1}. The resistance increases in much the same way as predicted, verifying that wires indeed reach the temperatures we expect.

At elevated temperatures, the enhanced diffusivity of atoms causes grains to grow rapidly, so the mean grain size will increase over time under the passage of an electric current. This is so significant that a

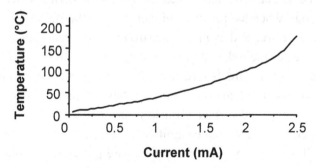

Fig. 5.25. Measured dependence of resistance and hence temperature on current for a 60 nm wide, 20 nm thick, 1 μm long gold nanowire on a 20 nm thick SiO$_2$ layer.

thin wire which has a size comparable to the mean grain size can, under current stressing, change from polycrystalline to bamboo at the centre of the wire (the point of highest temperature). This will cause a shift of the peak stress towards the centre of the wire to occur at a greater rate than is suggested by Fig. 5.10. This, coupled with the amount of surface diffusion which occurs for thin wires suggests that the thin wires will fail quite dramatically.

This trend of thin wires to fail dramatically has been observed experimentally, as shown in Fig. 5.26 below, which shows electron micrographs of wires of different widths after having failed.

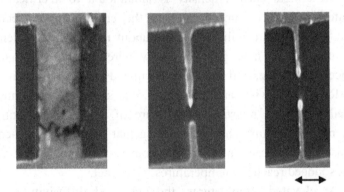

Fig. 5.26. SEM images of Au nanowires after failure, wire widths 320 nm, 60 nm and 40 nm. The scale bar is 200 nm.

The larger wire has failed along a meandering crack near the cathode end, whereas for the thinner wires, the failure has shifted towards the centre, and is more catastrophic. It even appears as if the thinnest wire has melted, and blown up, like a fuse. In Fig. 5.27 there are TEM images of a wire before and after current stressing and failure, where it is evident that the microstructure has completely changed.

The enhanced surface diffusion causes wires to thin down, so the local temperature can become significantly higher than predicted. As an example, if a 60 nm wide, 20 nm thick wire grown on 20 nm of oxide, and carrying 2 mA becomes thinned down to a diameter of 10 nm over a length of 25 nm, the peak temperature in that section could be as high as

Fig. 5.27. TEM images of a Au nanowire after failure, wire width 100 nm. The microstructure has been dramatically altered by the passage of current.

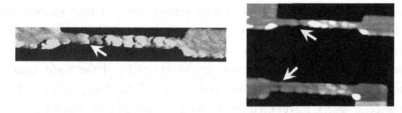

Fig. 5.28. AFM images of Au nanowires after failure. The arrows indicate the eventual failure positions.

1000°C (taking into account the increased resistivity and decreased thermal conductivity due to the small dimensions), which is close to the melting point of bulk gold. Nanostructured gold will melt at a slightly lower temperature than this, so the smallest wires will tend to fail in a catastrophic manner unless resistive heating can be kept to a minimum.

This can be accomplished by reducing the current through the wire to a low enough value to ensure that even when the wire thins down, the heating will never reach a high enough degree to cause the wire to fail catastrophically, so it will fail purely by diffusion. Some examples of this are shown in Fig. 5.28.

5.9. Consequences for Nanoelectronics

Heating and failure of interconnects is clearly a potentially significant problem facing the developers of nanoelectronic devices. At this point, it would be prudent to consider what types of future device incorporating such nanowires are most likely to be developed:

- Metallic interconnects of reduced dimensions in semiconductor microprocessor chips;
- Semiconductor nanowires;
- Metallic interconnects to either single or many molecule devices.

In the latter two cases, electromigration is hardly likely to be an issue, as the current levels tend to be extremely low in such devices, as there are far fewer charge carriers in semiconductors than metals, and molecular devices tend to have resistances in the $G\Omega$ range or higher. However, for conventional metallic interconnects, electromigration still poses a major problem, and there are many tricks of the trade used to extend their lifetimes, including passivation of the wire surfaces, dual-damascene layers, novel metallic alloys and reduced wire lengths.

For the remainder of this text, we will concern ourselves with molecular devices where the current levels are in the nA range, and electromigration and Joule heating in the interconnects need not be considered.

References for Chapter 5

1. J. E. Sanchez, L. Y. McKnelly and J. W. Morris, *J. Appl. Phys.* **72**, 3201 (1992).
2. J. E. Sanchez, O. Kraft and E. Arzt, *Appl. Phys. Lett.* **61**, 3121 (1992).
3. Y.-C. Joo and C. V. Thompson, *J. Appl. Phys.* **81**, 6062 (1997).
4. J.-M. Paik, I.-M. Park, Y.-C. Joo and K.-C. Park, *J. Appl. Phys.* **99**, 024509 (2006).
5. D. Josell, D. Wheeler and T. P. Moffat, *J. Electrochem. Soc.* **153**, C11 (2006).
6. J. R. Black, *IEEE Trans. Electron Dev.* **ED-16**, 338 (1969).
7. L. Berenbaum and R. Rosenberg, *Thin Solid Films* **4**, 187 (1969).

8. I. A. Blech, *J. Appl. Phys.* **47**, 1203 (1976) and subsequent erratum, I. A. Blech, *J. Appl. Phys.* **48**, 2648 (1977).

9. Y.-C. Joo and C. V. Thompson, *J. Appl. Phys.* **76**, 7339 (1994).

10. M. A. Korhonen, P. Borgesen, K. N. Tu and C.-Y. Li, *J. Appl. Phys.* **73**, 3790 (1993).

11. B. D. Knowlton, J. J. Clement and C. V. Thompson, *J. Appl. Phys.* **81**, 6073 (1997).

12. O. Kraft and E. Arzt, *Appl. Phys. Lett.* **66**, 2063 (1995).

13. B. D. Knowlton and C. V. Thompson, *J. Mater. Res.* **13**, 1164 (1998).

14. C. Durkan, M. A. Schneider and M. E. Welland, *J. Appl. Phys.* **86**, 1280 (1999).

15. C. Durkan and M. E. Welland, *Ultramicroscopy* **82**, 125 (2000).

16. M. Jakob, *Heat Transfer*, Chapter 11 (Wiley, 1949).

17. C. Kittel, *Introduction to Solid State Physics*, 6th edn. (Wiley, 1986).

18. *CRC Handbook of Chemistry and Physics*, 63rd edn. (CRC press, 1983).

19. C.-K. Hu, R. Rosenberg and K. Y. Lee, *Appl. Phys. Lett.* **74**, 2945 (1999).

20. J. Proost, L. Delaey, J. D'Haen and K. Maex, *J. Appl. Phys.* **91**, 9108 (2002).

Chapter 6

Elements of Single-Electron and
Molecular Electronics

In the previous chapters, we have seen how to describe current flow in structures with dimensions ranging from macroscopic to atomic. The mechanisms for current flow broadly fall into three classifications: drift-diffusion, ballistic, and tunnelling. There are other regimes of current flow which we will briefly consider in this chapter, namely single-electron transport (Coulomb blockade) and molecular transport, both of which can be understood in terms of what we have already considered.

6.1. Single-Electron Transport and Coulomb Blockade

Whenever an electron is added to a conductor, for instance when a current is flowing through any device, the additional charge will cause all of the already-present charges to re-arrange slightly. This has the effect of modifying the potential energy of the conductor. In most cases, this is not observable (except as shot noise), as the change is simply too small. However, when the conductor is small enough, this change can become appreciable and can have a dramatic effect on the transport characteristics.

Consider placing a small metal dot of capacitance C in between two tunnelling contacts as illustrated in Figs. 6.1 and 6.2. Under an applied bias between the contacts (as shown), an electron will tunnel from the left contact into the capacitor and then out into the right contact.

However, while the electron is in the capacitor, it adds an electrostatic charging energy of $e^2/2C$. If another electron tries to tunnel into the capacitor while the first one is still there, it can only do so when the bias is increased by $e/2C$. This effect is known as the Coulomb Blockade [1–3] which was predicted long before it was observed experimentally, and the device we have just considered is known as a single-electron transistor (SET).

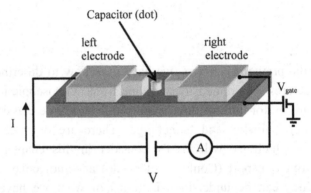

Fig. 6.1. Schematic of single-electron transistor configuration. A tunnel current, I flows through the device upon application of a voltage, V. This current will display Coulomb Blockade when the electrostatic charging energy of the capacitor is larger than a few k_BT. The current can also be modulated by applying a gate voltage, V_{gate}.

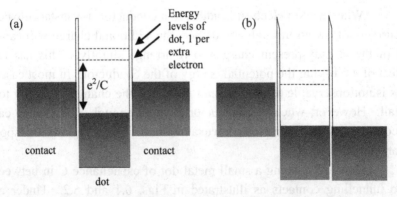

Fig. 6.2. Energy-level diagram for a device exhibiting Coulomb Blockade. (a) No applied bias: an energy gap exists, of magnitude e^2/C. Until the applied voltage reaches e/C, as in (b), no current will flow.

For this effect to be observable at room temperature (which is needed in order for such a device to be practical) the charging energy must be greater than the thermal energy, k_BT. To see what this entails, consider a metal sphere of radius r, surrounded by a medium with relative permittivity ε_r. To a good approximation, this will have a capacitance of $4\pi\varepsilon_o\varepsilon_r r$. In order to observe Coulomb Blockade, $e^2/2C > k_BT$, which at room temperature corresponds to $C \sim 10^{-18}$ F, which means that r must be less than about 5 nm for a metal particle on a SiO_2 substrate. Unfortunately, this is just below the resolution limit of electron-beam lithography, so there is currently no way to batch fabricate room temperature metal-based SETs in a top-down process. The situation is somewhat different for semiconductors, which have much larger capacitance for the same size as a conductor, due to their significantly lower electron density.

There are however, other means of fabricating SETs which were pioneered in the early 1990s, which incorporate top-down and bottom-up nanofabrication. The structure shown in Fig. 6.3(a) is a single-electron transistor formed by a chain of three 10 nm diameter gold colloids [4]. Figure 6.3(b) shows the dependence of drain current on source–drain voltage for various temperatures. Here we can see that the colloids are too big and hence have too large a capacitance to have an observable effect at room temperature, and Fig. 6.3(c) shows the dependence of drain current on gate voltage for various source–drain voltages, where we can see a gate effect. Therefore, SETs are an extreme form of FET.

We have seen that working SETs have been demonstrated, and some very interesting physics has been observed which has helped in our overall understanding of transport at the nanoscale. The questions at this point are (a) do SETs have a viable future as mass-produced devices, and (b) what are the uses of SETs? History tells us that we cannot predict the answer to the first question, and the uses of SETs are ever growing. By virtue of the way in which they operate, SETs are extremely sensitive electrometers, in fact, sensitive to around 10^{-5} e [5], so they are useful as electron detectors. However, until SETs capable of room temperature operation can be fabricated in a reliable way, this may be limited to one-off devices in research applications.

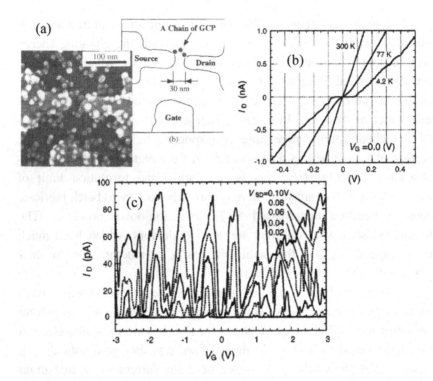

Fig. 6.3. (a) Single-electron transistor, (b) current–voltage characteristics as a function of temperature — by 300 K, the Coulomb Blockade is no longer measurable; (c) current–gate voltage characteristics showing characteristic conductance oscillations. From Ref. 4.

6.2. Molecular Electronics: Why Bother?

There has been a lot of interest in studying electronic transport through molecules in recent years. This intensive research has been undertaken for the following compelling reasons:

- Size — appropriate molecules can be very small, with nanometric dimensions. The smallest transistor that can be currently made has a volume of around $(20 \text{ nm})^3$, a factor of around 1000 times higher than the volume of a molecule.

This could ultimately lead to circuits with significantly higher integration than is currently possible.

- Speed — although the resistance of a molecule can be very high, the time taken for an electron to pass through one (the tunnelling time) is of the order 10^{-15} s. This is comparable to or less than the transit time for a transistor, which is of the order 10^{-14} s.

- Novel functionality — the mechanisms of current flow through nanoscale devices are fundamentally different than the larger devices we are used to. This leads to novel current–voltage characteristics which can be used to perform new functions. A related case in point is the single-electron transistor, which uses tunnelling and the Coulomb Blockade.

- Cost — the costs associated with semiconductor processing are increasing as the minimum feature size is decreasing. Each new process technology costs billions of dollars to implement, and consumers no longer accept large price increases just for a slightly faster computer or device. Therefore, the profit per transistor of the large semiconductor companies has started to decrease. Conceptually, then there is a lot to be gained if a transistor circuit could be *grown* on a silicon wafer, rather than fabricated using a process flow with hundreds of steps, each of which is expensive and time-consuming. The ultimate goal then of the molecular electronics community, apart from knowledge acquisition, is to be able to self-assemble entire circuits using tricks from organic chemistry. One can see the attractiveness of this idea: add a few drops of solution to a surface, and hey presto, you have a working circuit which has a better performance than a conventional circuit! This could be an extremely low-cost route to device fabrication. Each of the aspects of the above scenario are being actively researched, and it is not quite as fanciful as it may seem, although we are a long way from achieving these goals.

6.3. Mechanisms of Electron Transport Through Molecules

There are a variety of mechanisms based on tunnelling, thermionic emission, and hopping by which electrons can flow through molecules. In Table 6.1, we have briefly outlined each of these. Inelastic tunnelling also plays an important role, as molecular vibrations can assist in the tunnelling process, as we saw in Chapter 4.

Table 6.1. Conduction mechanisms through molecules.

Mechanism	Form of current	Voltage-dependence	Temperature-dependence
Low-voltage tunnelling	$J \sim V \dfrac{e^2 \alpha \sqrt{2m\varphi}}{h^2 s} e^{-2\alpha s \frac{\sqrt{2m\varphi}}{\hbar}}$	$J \propto V$	Very weak
Fowler–Nordheim (high voltage) tunnelling	$J \sim \dfrac{e}{4\pi^2 \hbar s^2} \left\{ \left(\varphi - \dfrac{eV}{2} \right) e^{-2\alpha s \frac{\sqrt{2m\left(\varphi - \frac{eV}{2}\right)}}{\hbar}} \right\}$	$\ln\left(\dfrac{J}{V^2}\right) \propto \dfrac{1}{V}$	Very weak
Hopping	$J \sim V e^{-\frac{\varphi}{k_B T}}$	$J \propto V$	$\ln\left(\dfrac{J}{V}\right) \propto \dfrac{1}{T}$
Thermionic emission	$J \sim T^2 e^{-\frac{\varphi - e\sqrt{\frac{eV}{4\pi\varepsilon_0 \varepsilon_r s}}}{k_B T}}$	$\ln(J) \propto \sqrt{V}$	$\ln\left(\dfrac{J}{T^2}\right) \propto \dfrac{1}{T}$

(where all the variables have their usual meaning, and T is temperature, rather than transmission coefficient)

It has been widely reported in the literature that the conductance of many families of molecules depends exponentially on the molecular length as follows

$$G = G_0 e^{-\beta L}, \qquad (6.1)$$

where β is the so-called decay constant within the molecule, and is given by

$$\beta_{LV} = \frac{2\sqrt{2m}}{\hbar} \alpha \sqrt{\varphi} \quad \text{for low bias and} \quad \beta_{HV} = \beta_{LV} \sqrt{1 - \frac{eV}{2\varphi}}$$

for higher bias (i.e. for $V > \varphi$). This length dependence is a characteristic of tunnelling. The factor α depends on the shape of the potential barrier, and on the electron effective mass within the molecule, and is known as the *barrier shape parameter*. A rectangular barrier with an effective mass of 1 has $\alpha = 1$.

We have already looked at tunnelling in detail in the earlier chapters, and found very little temperature dependence, except that at elevated temperatures, features in current–voltage characteristics tend to become smeared out.

Thermionic emission is a process which is always present, whereby electrons can be thermally excited over a potential barrier, as opposed to tunnelling through it. This clearly has a very strong temperature dependence, and will only become significant when the potential barrier is relatively small.

Hopping conduction is a process whereby a current becomes localised at certain points within a molecule, and can tunnel or hop between those points. This will also be a thermally activated process.

How can one determine the mechanism of conduction through a particular molecular device? It is clear from looking at Table 6.1 that it is not sufficient to analyse current–voltage characteristics alone, but their temperature dependence must also be measured. It is likely that several of these mechanisms will be present at the same time, so one can only hope to determine the dominant one. It is becoming standard practise to measure current–voltage characteristics at a variety of temperatures from 4.2 K up to room temperature and fit them to determine the average barrier height φ, the decay constant β, and the barrier shape parameter α.

There are situations when none of the above models are appropriate descriptions of the transport through a molecule. The most common deviation is that there may be several bands contributing to conduction at the same time. This will be the case for molecules containing HOMO and LUMO bands or levels, equivalent to the valence and conduction bands of a semiconductor. Under these circumstances, there will be holes as well as electrons contributing to conduction. To describe transport in this case, the Franz two-band model may be used [6–8].

6.4. Visualising Transport Through Molecules

Whilst the above information can give us some insight into conduction through molecules, it would nevertheless be instructive to consider *how* electrons travel through molecules. We saw in Chapter 1 that conduction through metals is relatively straightforward: electronic wave-functions can extend throughout a conductor but electrons get scattered from defects and imperfections in the lattice. On the whole, electrons within a conductor are *delocalised*. The picture is very different in molecules, where the electrons can be localised in some regions and delocalised in others, hence the relevance of the hopping mechanism of conduction. What gives rise to delocalisation or localisation in molecules? From a simple standpoint, it depends on whether the electrons are in conjugated or non-conjugated bonds, respectively, i.e. π or σ bonds, as illustrated in Fig. 6.4.

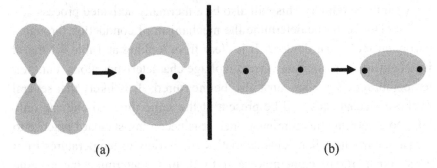

(a)   (b)

Fig. 6.4. (a) π-bonds, within which electrons are delocalised; (b) σ-bonds, within which electrons are localised.

When dealing with an entire molecule, there are sophisticated ways based on density functional theory (DFT) [9,10], of calculating the spatial and energy distribution of the electron wave-functions. This can then be used to calculate current–voltage characteristics of the molecules in a device configuration, for comparison with experiment. An example is shown in Fig. 6.5.

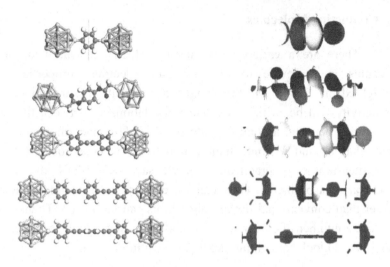

Fig. 6.5. Left: schematic of five different molecules bridging gaps between gold electrodes, and right: HOMO levels of molecules. As the molecules increase in length, and the distance between ring structures increases, the electron wave-functions become more localised, with the result of reduced conductance. From Ref. 11.

6.5. The Contact Resistance Problem

Initial experiments on trying to measure transport through molecules concentrated on long molecules, such as DNA [12] and single-walled carbon nanotubes [13]. Reports on resistance measurements for both were wildly conflicting, in many cases carrying by several orders of magnitude for nominally similar molecules. The trends were quite clear: higher resistances were usually measured when molecules were placed on top of electrodes than when the electrodes were placed on top of the molecules, although this is not always the case. Subsequent work has shown that in the former case, the contact resistance is significantly higher than in the latter case. When depositing electrodes on top of molecules, one must be careful that the evaporated metal atoms do not have so much kinetic energy that they damage the molecule. Work on this topic is ongoing, but the standard approach is now to deposit electrodes on top of molecules whenever possible.

6.6. Contacting Molecules

There are a variety of means by which one can incorporate molecules in a device to measure their electrical properties. As highlighted in Eq. (6.1), the longer a molecule is, the lower its conductivity will be, with an exponential relationship between the two. As β varies between around 0.6 and 0.95 Å^{-1} [14] once molecules are longer than around 2–3 nm, their conductivity becomes too low to be either measurable or practical.   As we saw with SETs, this poses somewhat of a problem. How can we connect to something which is smaller than conventional lithographic limits can fabricate? Thankfully, there are a number of clever ways around this. In principle, a molecular device should look something like that shown in Fig. 6.6.

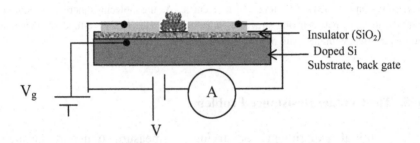

Fig. 6.6. Schematic of molecular electronic device.

How should one go about connecting molecules? There are a number of ways of forming nanogap devices that have been used with varying degrees of success during recent years, as follows:

- Electron-beam lithography;
- Electromigration;
- Mechanically-controlled break junctions (MCBJ);
- Molecular sandwiches;
- STM.

We have already encountered all of these concepts, apart from MCBJs, and we shall now consider them again within the context of molecular electronics experiments.

6.6.1. *Nanogaps formed by electron-beam lithography*

As we saw earlier, e-beam lithography cannot reproducibly form structures with a resolution below around 5 nm, and for molecular electronics experiments, we need electrodes as close together as around 1–2 nm. The key here is in the word *reproducibly*, because if we try to fabricate electrodes with a given separation, we will end up with a distribution of separations. A technique, pioneered by M. S. M. Saifullah has produced a yield of coplanar electrodes with 75% having sub-5 nm gaps [15]. Some examples are shown in Fig. 6.7.

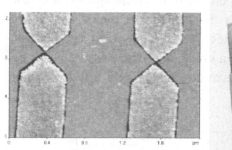 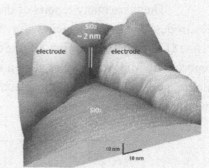

Fig. 6.7. AFM images of coplanar nanogaps formed by e-beam lithography. Gaps were imaged using a carbon-nanotube AFM tip [15].

6.6.2. *Nanogaps formed by electromigration*

As we have already seen in Chapter 5, when a current flows through a wire, the electrons cause material to drift along the length of the wire, eventually leading to failure. If a wire is left to fail slowly by electromigration under static conditions, the failure tends to be catastrophic, as shown in Fig. 5.26. However, if the current through the

Fig. 6.8. SEM images of a 15 nm wide, 20 nm thick Au nanowire which has been failed by electromigration to produce a very thin crack (see arrow). The wire is decorated with residual e-beam resist which does not interfere with electrical measurements [author's own unpublished results].

wire is ramped up at a suitable rate a number of times, so that electromigration effectively proceeds in pulses, the failure can be carefully controlled, and the wire can be made to fail along a crack as narrow as ~ 1 nm. An example is shown in Fig. 5.28. Another example is shown in Fig. 6.8.

There are many reports of the use of this technique to form nano-electrodes with nm spacing for molecular electronics experiments [16–18]. One of the first of these was the so-called C_{60} transistor reported by McEuen [18], as shown in Fig. 6.9. The evidence points to a single molecule between the electrodes, and a gate effect is clearly measurable.

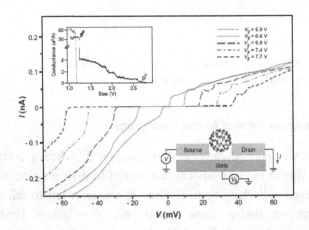

Fig. 6.9. C_{60} transistor characteristics as reported in Ref. 18.

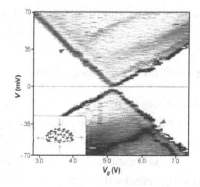

Fig. 6.10. Conductance map of C_{60} transistor as reported in Ref. 18. The arrows point to bands at 35 meV, corresponding to the vibrational mode shown in the inset.

The conductance map shown in Fig. 6.10 shows evidence that the vibrational modes of C_{60} play a role in the transport.

Whilst some important breakthroughs in the field of molecular electronics have been made using this configuration, there are a number of issues which pose difficulties:

- Reproducibility in the formation of nanogaps;
- Unintended incorporation of metallic nanoparticles in gap leading to observation of Coulomb Blockade, rendering devices useless for molecular electronics experiments;
- Lack of control over nanogap separation once formed.

The first point has been addressed by Strachan [19] who has reported a yield of almost 100% of useful devices. The second point seems to only be a particular problem in the case of wires which are failed at low temperatures. The most limiting problem however is the last one. Suppose we want to perform a measurement on a molecule which has a nominal length of 1 nm. If the nanogap is larger than this, say 1.1 nm, then a molecule will not be able to span the gap, and instead transport may be through several molecules in series. If the gap is smaller than 1 nm, then molecules will be able to span the gap if they are not exactly parallel to the wire. The optimum is to have the gap exactly the same as the molecular length. This is something which will only occur by chance with electromigration-formed nanogaps, hence the development of break junctions which we will consider next.

6.6.3. *Mechanically-controlled break junctions*

Several beautiful key experiments have been performed during recent years using mechanically controlled break-junctions (MCBs) which demonstrate transport measurements through single (or very few) molecules [20–22]. Initial experiments with break junctions were initiated after interesting conduction properties were observed with atomic point contacts [23–25]. These were generally made by pulling apart a thin wire in a highly controlled manner, whilst observing it in a high resolution TEM, and whilst measuring its electrical conductance. Using MCBs it is possible not only to form a nanogap, but to then repeatedly alter its size to be commensurate with the molecules of interest. The way a typical MCB molecular electronics experiment progresses now has several steps, as illustrated in Fig. 6.11:

1. Lithographically pattern metal electrodes on a flexible, insulating substrate. Typically use a thin (~1 μm) layer of polyimide on a metal foil;
2. Bend substrate (using fine mechanical control) to cause electrodes to break at a pre-defined weak point which has nanometric dimensions. Calibrate electrical characteristics of junction — should be a measurable tunnel current if gap is below 3–4 nm. Ensure gap is larger than length of molecules;
3. Add drop of solution containing molecules;
4. Close gap until a jump in conductance indicates that a molecule has bridged the gap. Perform transport measurements;
5. Repeatedly break and re-form device, to assess reliability;
6. Compare measurements with DFT modelling.

In this way, an ever-increasing number of molecular systems have been studied in detail within recent years, including Benzene dithiol,

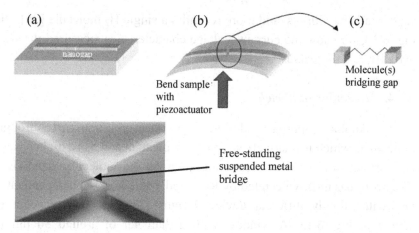

Fig. 6.11. Schematic of MCBJ experiments. A wire is formed and then broken along a nm-sized crack. The spacing between the resulting nanoelectrodes may be changed by bending the substrate.

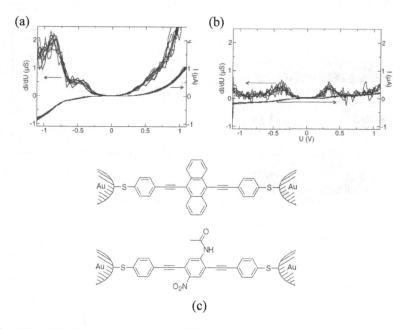

Fig. 6.12. *I–V* characteristics for two different molecules measured using a MCBJ. (a) Symmetric molecule, showing symmetric characteristics, (b) asymmetric molecule showing asymmetric characteristics, (c) symmetric (top) and asymmetric (bottom) molecules used in experiments [22].

various alkane dithiols, and more recently, a single H_2 molecule [26]. In Fig. 6.12, we show the current–voltage characteristics measured through two different molecules.

6.6.4. *Molecular sandwiches*

Another approach designed to measure transport through molecules, which is relatively simple to implement and has a higher yield of working devices than electromigration, is the molecular sandwich. This work has mostly been led by Reed and Tour, working independently and with slightly different device designs [27–29]. The concept is shown in Fig. 6.13. A window with a diameter of around 50 nm is lithographically formed in a Si_3N_4 membrane. Au is then evaporated onto one side of the membrane, and this forms one of the contacts. A SAM is then deposited on the Au within the window, and finally, a layer

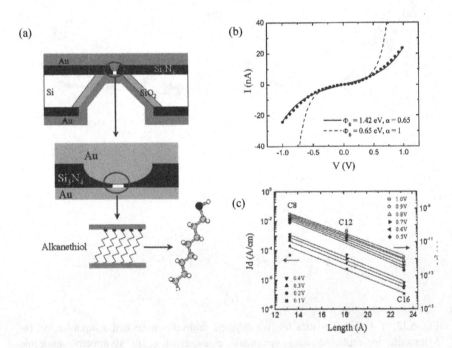

Fig. 6.13. Molecular sandwich device, (a) schematic, (b) *I–V* measurement, (c) scaling of current with molecular length [27].

of Au is deposited on top of the SAM. In this way, the SAM is sandwiched between two Au electrodes, and good electrical contact is achieved through the Au-thiol bond of the SAM. In these devices, the current is flowing through many thousands of molecules rather than through a single one, with the consequence of increased current levels and reduced noise.

The results obtained through this approach as well as with MCBJs have enabled the field of molecular electronics to progress rapidly during recent years. There are systematic studies on the effect on conductance of altering the chain length and side-groups of various types of molecule. It has been experimentally verified that (a) long molecules do indeed have a lower conductance than shorter molecules, (b) conjugated molecules have a higher conductance than non-conjugated molecules, and (c) molecules with a structural asymmetry exhibit asymmetric current–voltage characteristics.

6.6.5. STM probing of molecules

We briefly encountered this topic in Chapter 4 where we introduced the STM. This technique has a lot to offer the field of molecular electronics, as it opens the way to perform experiments on single molecules under highly controlled conditions. Whilst this is not a scaleable technology, it does allow us to gain a better understanding of transport through molecules, which can then be applied in real devices. An example of a current–voltage characteristics and image of molecules obtained by STM is shown in Fig. 6.14.

By means of such STM-based measurements, it is possible to extract information regarding the density of states of the molecules and the decay coefficient, β, as shown in Fig. 6.15 [30].

There are many thousands of reports in the literature of STM measurements on molecules, with a wide range of transport characteristics being measured, including resonant tunnelling and rectification. For a good overview, the reader is referred to Ref. 31.

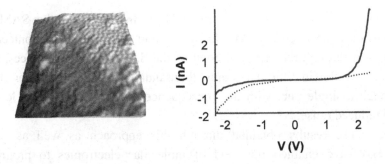

Fig. 6.14. STM probing of Lander molecules on Si(111). (a) Topography, (b) current–voltage characteristics: solid and dotted curves are on molecule and Si, respectively. Author's own, unpublished work.

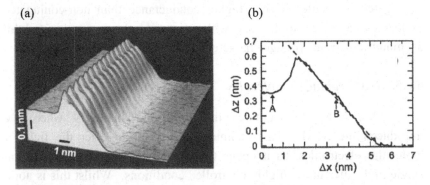

Fig. 6.15. STM probing of "Lander" molecules on a Cu (100) surface. (a) Topography (constant density of states surface) of an array of 15 molecules aligned at a double step edge, (b) apparent molecular height along a molecule, from which a decay coefficient of 4 nm^{-1} was inferred. From Ref. 30.

6.7. The Future

What does the future hold for single-electron and molecular electronics? As before, one can only speculate about this, while remembering that already a considerable amount of effort and resources have been poured into investigating both. Already as we have seen, the applications for single-electron transistors are steadily growing. The picture is less clear for molecular devices, mainly due to issues regarding reproducibility, reliability and yield. No doubt research into each of these aspects will continue apace, and some day molecular devices may be commonplace in classical or quantum computers. It must also be remembered that (low-temperature) single-electron transistors could be fabricated using conventional means, and had a head-start of a few years on molecular devices.

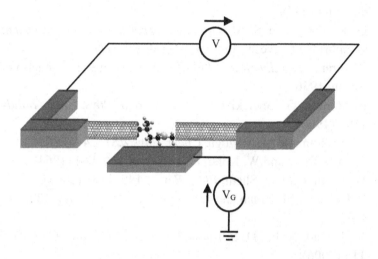

Fig. 6.16. Schematic of possible future molecular electronic device, where the molecule is contacted using single-walled carbon nanotubes. The device has an individual gate electrode.

Just thinking about the research devices of today, the biggest single problem is in the formation of stable and reliable electrical contacts to a molecule. One might imagine that this may be overcome if circuits can be grown or synthesised, contacts included. A particularly elegant concept for a future device is illustrated in Fig. 6.16. Here, a molecule is shown which is attached to two single-walled carbon nanotubes, on which electrodes are deposited. Metallic carbon nanotubes are excellent conductors of electricity, but as yet there is no way to preferentially synthesise conducting ones (only 1/3 of carbon nanotubes are conducting, the rest are semiconducting).

References for Chapter 6

1. S. Datta, *Electronic Transport in Mesoscopic Systems* (Cambridge University Press, Cambridge, 1995).
2. C. J. Gorter, *Physica* **17**, 777 (1951).
3. C. W. J. Beenakker, *Phys. Rev. B* **44**, 1646 (1991).
4. T. Sato, H. Ahmed, D. Brown and B. F. G. Johnson, *J. Appl. Phys.* **82**, 696 (1997).
5. D. K. Ferry and S. M. Goodnick, *Transport in Nanosctructures* (Cambridge Univesity Press, Cambridge, 1997).
6. W. Franz, *Handbuch der Physik*, ed. S. Flugge (Springer-Verlag, Berlin, 1956).
7. R. Stratton, G. Lewicki and C. A. Mead, *J. Phys. Chem. Solids* **27**, 1599 (1966).
8. C. Joachim and M. Magoga, *Chem. Phys.* **281**, 247 (2002).
9. P. Hohenberg and W. Kohn, *Phys. Rev. B* **136**, 864 (1964).
10. W. Kohn and L. J. Sham, *Phys. Rev. A* **140**, 1133 (1965).
11. P. Delaney, M. Nolan and J. C. Greer, *J. Chem. Phys.* **122**, 044710 (2005).
12. X.-T. Gao, X. Fu, D.-S. Liu and S.-J. Xie, *J. Phys.: Conf. Ser.* **29**, 115 (2006).
13. T. W. Ebbesen, H. J. Lezec, H. Hiura, J. W. Bennett, H. F. Ghaemi and T. Thio, *Nature* **382**, 54 (1996).
14. X. D. Cui, X. Zarate, J. Tomfohr, O. F. Sankey, A. Primak, A. L. Moore, T. A. Moore, D. Gust, G. Harris and S. M. Lindsay, *Nanotechnology* **13**, 5 (2002).
15. M. S. M. Saifullah, T. Ondarcuhu, D. K. Koltsov, C. Joachim and M. E. Welland, *Nanotechnology* **13**, 659 (2002).
16. S. Ghosh, H. Halimun, A. K. Mahapatro, J. Choi, S. Lodha and D. Janes, *Appl. Phys. Lett.* **87**, 233509 (2005).
17. M. Austin and S. Y. Chou, *J. Vac. Sci. Technol. B* **20**, 665 (2002).
18. H. Park, A. J. L. Lim, A. P. Alivisatos, J. Park and P. L. McEuen, *Appl. Phys. Lett.* **75**, 301 (1999).
19. D. R. Strachan, D. E. Smith, D. E. Johnston, T. H. Park, M. J. Therein, D. A. Bonnell and A. T. Johnson, cond-mat/0504112, 5 April 2005.

20. M. A. Reed, C. Zhou, C. J. Muller, T. P. Burgin and J. M. Tour, *Science* **278**, 252 (1997).

21. J. Reichert, H. B. Weber, M. Mayor and H. V. Lohneysen, *Appl. Phys. Lett.* **82**, 4137 (2003).

22. J. Reichert, R. Ochs, D. Beckmann, H. B. Weber, M. Mayor and H. V. Lohneysen, *Phys. Rev. Lett.* **88**, 176804-1 (2002).

23. G. Rubio, N. Agrait and S. Vieira, *Phys. Rev. Lett.* **76**, 2302 (1996).

24. Y. Kondo and K. Takayanagi, *Science* **289**, 606 (2000).

25. A. M. C. Valkering, A. I. Mares, C. Untiedt, K. B. Gavan, T. H. Oosterkamp and J. M. van Ruitenbeek, *Rev. Sci. Instrum.* **76**, 103903 (2005).

26. D. Djukic, K. S. Thygssen, C. Untiedt, R. H. M. Smit, K. W. Jacobsen and J. M. van Ruitenbeek, *Phys. Rev. B* **71**, 161402-1 (2005).

27. W. Wang, T. Lee and M. A. Reed, *Phys. Rev. B* **68**, 035416 (2003).

28. C. Zhou, M. R. Deshpande, M. A. Reed, L. Jones, II and J. M. Tour, *Appl. Phys. Lett.* **71**, 611 (1997).

29. W. Wang, T. Lee, M. Kamdar, M. A. Reed, M. P. Stewart, J.-J. Hwang and J. M. Tour, *Supperlatt. Microstruc.* **33**, 217 (2003).

30. V. J. Langlais, R. R. Schlittler, H. Tang, A. Gourdon, C. Joachim and J. K. Gimzewski, *Phys. Rev. Lett.* **83**, 2809 (1999).

31. J. K. Gimzewski and C. Joachim, *Science* **283**, 1683 (1999).

Solutions to Problems in Chapter 2

1. **Ball**:
   deBroglie wavelength, $\lambda = h/mv = 6.6 \times 10^{-34}/(10^{-3} \times 150) = 4.4 \times 10^{-33}$ m. Energy $= (1/2)\ mv^2 = 11.25$ J $= 7.031 \times 10^{19}$ eV. Planck frequency $= E/h = 1.706*10^{34}$ Hz.

   Car:
   deBroglie wavelength, $\lambda = h/mv = 4.4 \times 10^{-42}$ m. Energy $= (1/2)\ mv^2 = 4.05 \times 10^{12}$ J $= 2.531 \times 10^{31}$ eV. Planck frequency $= E/h = 6.136*10^{45}$ Hz.

   O_2 molecule:
   Mass of molecule $= 32$ g/Avagadro's constant (i.e. number of molecules per mole) $= 32 \times 10^{-3}/6.02 \times 10^{23} = 5.3*10^{-26}$ kg
   deBroglie wavelength, $\lambda = h/mv = 1.242$ nm. Energy $= (1/2)\ mv^2 = 2.66 \times 10^{-24}$ J $= 1.66 \times 10^{-5}$ eV. Planck frequency $= E/h = 4.03$ GHz.
   The molecule's behaviour will be dictated by quantum mechanics, as its deBroglie wavelength is comparable to the molecular dimensions.

2. $R = 0.016$. For length $= 2$ nm, $R = 0.01$
3. $R = 0.269$
4. c
5. Assume that well is infinitely deep and square. Therefore, we can write energy eigenvalues as:
   $E_n = \dfrac{h^2 n^2}{8ml^2}$ where l is the width of the well, and m is the electron mass within the well.

For the valence band, then, $E_{h1} = 0.05$ eV, and for the conduction band $E_{e1} = 0.25$ eV

⇨ emission wavelength of laser = 1.2 eV + 0.05 eV + 0.25 eV = 1.5 eV.

1.5 eV = 2.4×10^{-19} J = $hc/\lambda \Rightarrow \lambda = 825$ nm.

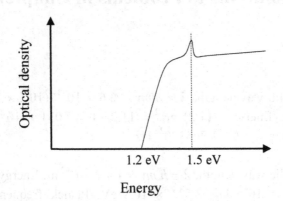

Index